United States
Department
of Agriculture

Forest Service

**Rocky Mountain
Research Station**

General Technical
Report RMRS-GTR-42-
volume 5

December 2002

Wildland Fire in Ecosystems

I0491278

Effects of Fire on Air

Abstract

Sandberg, David V.; Ottmar, Roger D.; Peterson, Janice L.; Core, John. 2002. **Wildland fire on ecosystems: effects of fire on air**. Gen. Tech. Rep. RMRS-GTR-42-vol. 5. Ogden, UT: U.S. Department of Agriculture, Forest Service, Rocky Mountain Research Station. 79 p.

This state-of-knowledge review about the effects of fire on air quality can assist land, fire, and air resource managers with fire and smoke planning, and their efforts to explain to others the science behind fire-related program policies and practices to improve air quality. Chapter topics include air quality regulations and fire; characterization of emissions from fire; the transport, dispersion, and modeling of fire emissions; atmospheric and plume chemistry; air quality impacts of fire; social consequences of air quality impacts; and recommendations for future research.

Keywords: smoke, air quality, fire effects, smoke management, prescribed fire, wildland fire, wildfire, biomass emissions, smoke dispersion

The volumes in "The Rainbow Series" will be published through 2003. The larger bold check-mark boxes indicate the volumes currently published. To order, check any box or boxes below, fill in the address form, and send to the mailing address listed below. Or send your order and your address in mailing label form to one of the other listed media.

☐ RMRS-GTR-42-vol. 1. **Wildland fire in ecosystems: effects of fire on fauna.**

☐ RMRS-GTR-42-vol. 2. **Wildland fire in ecosystems: effects of fire on flora.**

☐ RMRS-GTR-42-vol. 3. **Wildland fire in ecosystems: effects of fire on cultural resources and archeology.**

☐ RMRS-GTR-42-vol. 4. **Wildland fire in ecosystems: effects of fire on soil and water.**

☐ RMRS-GTR-42-vol. 5. **Wildland fire in ecosystems: effects of fire on air.**

Send to: _____
Name

_____
Address

Fort Collins Service Center	
Telephone	(970) 498-1392
FAX	(970) 498-1396
E-mail	rschneider/rmrs@fs.fed.us
Web site	http://www.fs.fed.us/rm
Mailing Address	Publications Distribution Rocky Mountain Research Station 240 W. Prospect Road Fort Collins, CO 80526-2098

Wildland Fire in Ecosystems

Effects of Fire on Air

Authors

David V. Sandberg, Research Physical Scientist, Corvallis Forestry Sciences Laboratory, Pacific Northwest Research Station, U.S. Department of Agriculture, Corvallis, OR 97331

Roger D. Ottmar, Research Forester, Seattle Forestry Sciences Laboratory, Pacific Northwest Research Station, U.S. Department of Agriculture, Seattle, WA 98103

Janice L. Peterson, Air Resource Specialist, Mt. Baker-Snoqualmie National Forest, U.S. Department of Agriculture, Mountlake Terrace, WA 98053

John Core, Consultant, Core Environmental Consulting, Portland, OR 97229

Cover photo—Photo by Roger Ottmar. Smoke blots out the sun during the 1994 Anne Wildfire in western Montana.

Preface

In 1978, a national workshop on fire effects in Denver, Colorado, provided the impetus for the "Effects of Wildland Fire on Ecosystems" series. Recognizing that knowledge of fire was needed for land management planning, state-of-the-knowledge reviews were produced that became known as the "Rainbow Series." The series consisted of six publications, each with a different colored cover, describing the effects of fire on soil, water, air, flora, fauna, and fuels.

The Rainbow Series proved popular in providing fire effects information for professionals, students, and others. Printed supplies eventually ran out, but knowledge of fire effects continued to grow. To meet the continuing demand for summaries of fire effects knowledge, the interagency National Wildfire Coordinating Group asked Forest Service research leaders to update and revise the series. To fulfill this request, a meeting for organizing the revision was held January 4-6, 1993, in Scottsdale, Arizona. The series name was then changed to "The Rainbow Series." The five-volume series covers air, soil and water, fauna, flora and fuels, and cultural resources.

The Rainbow Series emphasizes principles and processes rather than serving as a summary of all that is known. The five volumes, taken together, provide a wealth of information and examples to advance understanding of basic concepts regarding fire effects in the United States and Canada. As conceptual background, they provide technical support to fire and resource managers for carrying out interdisciplinary planning, which is essential to managing wildlands in an ecosystem context. Planners and managers will find the series helpful in many aspects of ecosystem-based management, but they will also need to seek out and synthesize more detailed information to resolve specific management questions.

— The Authors
December 2002

Acknowledgments

The Rainbow Series was compiled under the sponsorship of the Joint Fire Science Program, a cooperative fire science effort of the U.S. Department of Agriculture, Forest Service, and the U.S. Department of the Interior, Bureau of Indian Affairs, Bureau of Land Management, Fish and Wildlife Service, National Park Service, and U.S. Geological Survey.

Several scientists provided significant input without requesting authorship in this volume. We acknowledge valuable contributions by Sue A. Ferguson, Timothy E. Reinhardt, Robert Yokelson, Dale Wade, and Gary Achtemeier. We also thank the following individuals for their suggestions, information, and assistance that led to substantial technical and editorial improvements in the manuscripts: Scott Goodrick, Allen R. Riebau, Sue A. Ferguson, and Patti Hirami. Finally, we appreciate Marcia Patton-Mallory and Louise Kingsbury for persistence and support.

Contents

Summary

Wildland fire is an integral part of ecosystem management and is essential in maintaining functional ecosystems, but air pollutants emitted from those fires can be harmful to human health and welfare. Because of the public and governmental concerns about the possible risk of wildland fire smoke to public health and safety, as well as nuisance, visibility, ozone generation, and regional haze impacts, increasingly effective smoke management programs and air quality policies are being implemented with support from research and land management agency programs.

This state-of-knowledge review of what is known about the effects of fire on air quality has been prepared to assist those in the fire and air quality management communities for future discussion of management, policy, and science options for managing fire and air quality. The introduction sets up a framework in which to discuss the interaction between pollutants emitted from fire, and air quality at the national, State, and local levels applied to air resource management, fire management, and geographical scale components. It also provides an overview of science reviews conducted since 1979 and discusses recent changes in fire policy, strategies, and funding. The Clean Air Act and its amendments are discussed in chapter 2, in the context of how and why fire impacts each issue, what information is needed, and who needs it to fulfill legal requirements under the act. National ambient air quality standards, regional haze and visibility, hazardous air pollutants, and best available control methods are some of the topics covered. Chapter 3 covers the magnitude of the impacts of prescribed and wildland fire on air quality, and contains an overview of smoke management plans intended to manage those impacts.

Chapters 4 through 7 present scientific and technical discussions. Chapter 4 discusses the characterization and production rate of emissions from fire in terms of fuels, fire behavior, stages of combustion, fuel consumption, and emission factors of various pollutants. The basic elements and modeling of transport and dispersion are covered in chapter 5, including, plume, puff, particle, and grid models. Chapter 6 considers plume and atmospheric chemistry, the chemical reactions that occur in plumes, with a focus on ozone formation and particle formation. Use of emission inventories, air quality monitoring, and source apportionment methods, and mechanistic models to estimate the impacts of fire on air quality are covered in chapter 7. Chapter 8 reviews the health, welfare, economic, and safety consequences of these impacts. The final chapter recommends priorities for future research to better understand and quantify fire and its effect on air quality.

Metric Equivalents

When you know:	Divide by:	To find:
Feet (ft)	3.28	Meters
Pounds (lb)	2.21	Kilograms
Acres	2.47	Hectares
Pounds per acre	0.89	Kilograms per hectare
Fahrenheit (°F)	1.8 and subtract 32	Celsius

Chapter 1:
Introduction

A state-of-knowledge review, *Effects of Fire on Air*, was written in 1979 to inform environmental agencies, fire managers, and land management planners, and to guide research strategies in the intervening years (Sandberg and others 1979). That review is still technically sound for the most part, but substantial new knowledge is now available. In this volume, we update that review of knowledge important for managing the effects of fire on air and for adjusting the course of new research. In addition, we expand the scope of our review to place the information in the context of new policies regarding fire management and air quality management

Acquisition of scientific knowledge regarding air pollution from fires is motivated by active policy development both to restore the role of fire in ecosystems and to improve air quality. Land managers require quantitative analysis and goal-seeking solutions to minimize the negative consequences of fire management. Managing fire and air quality to the standards set by Congress requires an increasingly detailed base of scientific knowledge and information systems.

The *Federal Wildland Fire Policy* (U.S. Department of the Interior and U.S. Department of Agriculture 1995) and the *Clean Air Act as Amended 1990* (PL 101-549) resulted in the need to significantly raise the level of knowledge about fire's effects on air in order to meet regulatory and management requirements. For example, new information is needed to assess, monitor, predict, and manage:

- Emissions and air quality impacts from wildfires
- Acute health effects of human exposure to smoke
- Natural and anthropogenic sources of visibility reduction
- Cumulative air quality impacts from expanded fuel management programs
- Tradeoffs between air quality impacts from wildland fire and prescribed fire

Likewise, management of fire and air quality is also undergoing substantial policy development that has led to the need for new and different information to satisfy regulatory and management requirements. As both legal and management issues mature, there is less a sense that environmental regulation is a limitation on fire management, and more of a sense that ecosystem management goals, fire safety, and air quality are goals to be met collectively. For example, new air quality rules recognize the importance of the role of fire in sustaining ecosystems and the inherent tradeoffs between prescribed fire and wildland fire occurrence. At the same time, land management plans and real-time fire management decisions increasingly factor in the expected consequences to air quality.

USDA Forest Service Gen. Tech. Rep. RMRS-GTR-42-vol. 5. 2002

1

Since 1995, researchers and land managers have concentrated a great deal of energy to extend what is known about fire and its effect on air quality; to expand information systems that make knowledge readily available to policy, management, and public clients; to merge what is known about sustainable ecosystems and disturbance ecology with what is known about the chemistry, physics, biology, and social impacts of air pollution; and to redefine the research agenda.

Objective

This review summarizes the current state of knowledge of the effects of fire on air, and defines research questions of high priority for the management of smoke from fires. We also intend this as a reference document for future discussion of management, policy, and science options for managing fires and air quality. This review is limited to readily available published and unpublished knowledge and to original contributions by the authors. No new analysis of data or policy, nor assessment of impacts and options, is included herein.

Related Publications

This document does not stand alone. There are several excellent sources for information on the effects of fire on air. We advise the reader to include at least the following publications, each of which will be abstracted in this document, in your reference library:

- *Smoke Management Guide for Prescribed and Wildland Fire: 2001 Edition* (Hardy and others 2001)
- *National Strategic Plan: Modeling and Data Systems for Wildland Fire and Air Quality* (Sandberg and others 1999)
- *Introduction to Visibility* (Malm 2000)
- *Fire Effects on Air* (Sandberg and others 1979)
- *Southern Forestry Smoke Management Guidebook* (Southern Forest Fire Laboratory Personnel 1976)
- *Development of Emissions Inventory Methods for Wildland Fire* (Battye and Battye 2002)

Why, then, is another state-of-knowledge review necessary on the subject of fire effects on air? First, because policy and regulatory development in air quality management and in fire management is advancing rapidly, and there is a continuing need to reassess current knowledge about what is required to meet new expectations. Second, this document addresses the advancement of science at a much higher level than the above-mentioned references. Third, because the Joint Fire Science Program has sponsored a series of reviews, nicknamed the Rainbow Series (see

"Preface"), to compile a broad reference of fire effects to serve practitioners and policymakers charged with using and managing fire, and this is the third volume in that series. Finally, we hope you will find this volume a useful attempt to abstract and fill in the gaps left by the previous publications.

Scope

This review includes all health and welfare effects of air pollution from fires, but does not include the effects of air resource management on ecosystem health or any other value. Unless otherwise specifically stated, the term "fires" in this manuscript includes all prescribed and wildland fires on wildlands. Prescribed fires are ignited intentionally to achieve ecosystem management or fire protection objectives, whereas wildland fires result from unplanned ignitions on wildlands. Wildlands include all the nonagricultural and nonresidential rural lands of the United States, including the wildland-urban interface, regardless of ownership, sovereignty, or management objective. Management response to wildland fires differs greatly according to economic efficiency, the values at risk (including air quality), and the expected ecological consequences. Wildfires are at one end of the spectrum of wildland fires in that they are unwanted and unplanned, and are managed to minimize cost plus loss. At the other end of the spectrum are wildland fires that benefit ecosystem values, and are managed to maximize their benefit. Ideally, each wildland fire is evaluated with respect to expected costs, losses, risks, and benefits in order to provide an appropriate and preplanned response. Because fires are a significant emitter of air pollutants, many other fire management activities such as fire prevention or fuel treatment may have an indirect effect on air quality.

Framework

The issues, responsibilities, and tools that address fire and air quality are varied and complex, sometimes resulting in confusion about the physical scale and temporal stage of three characteristics: the application to fire management, the application to air resource management, and the physical process of air pollution. *National Strategic Plan: Modeling and Data Systems for Wildland Fire and Air Quality* (Sandberg and others 1999) provides a conceptual framework for visualizing fire's effects on air by representing the scope of the problem as a three dimensional array of air resource management, fire management, and scale components (fig. 1-1). The air resource component is ordered in time from emissions source strength, to ambient air quality, and to effects. The fire management component includes planning, operations, and

2

USDA Forest Service Gen. Tech. Rep. RMRS-GTR-42-vol. 5. 2002

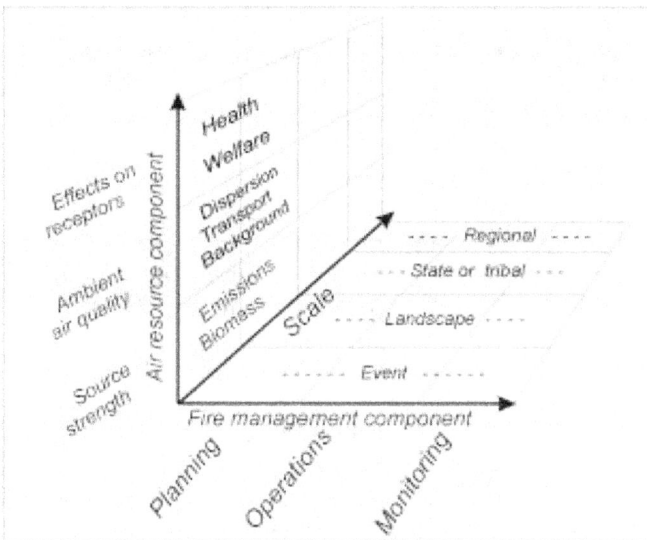

Figure 1-1—Three primary components of the issues, responsibilities, and tools related to wildland fire and air quality: air resource management, fire management, and scale (Sandberg and others 1999).

monitoring. The scale component includes the event, landscape, state or tribal, and regional scales.

We have organized this volume around the air resource component and expanded it to include a regulatory perspective (fig. 1-2). Fire in the context of the regulatory environment is the subject of chapters 2 and 3. Biomass consumption and emissions are the subject of chapter 4; transport and dispersion of pollutants in the atmosphere the subjects of chapters 5 and 6; air quality impacts the subject of chapter 7; and the effect on human values from exposure to air pollutants the subject of chapter 8. We conclude with a review of recommendations for future research in chapter 9.

Prior Work

Since the publication of *Effects of Fire on Air* (Sandberg and others 1979), significant changes have come to pass in both the technical and policy issues that surround the fire and air quality dilemma. The conferences, stakeholder group discussions, and technical publications discussed here have helped to shape the current fire management programs and will influence future programs.

Smoke Management Guide For Prescribed and Wildland Fire: 2001 Edition

Smoke Management Guide for Prescribed and Wildland Fire: 2001 Edition (Hardy and others 2001) has been developed by the Fire Use Working Team of the National Wildfire Coordinating Group (NWCG) and involves most of the same authors as this current publication. The guide provides fire management and smoke management practitioners with a fundamental understanding of fire emissions processes and impacts, regulatory objectives, and tools for the management of smoke from fires. It is a comprehen-

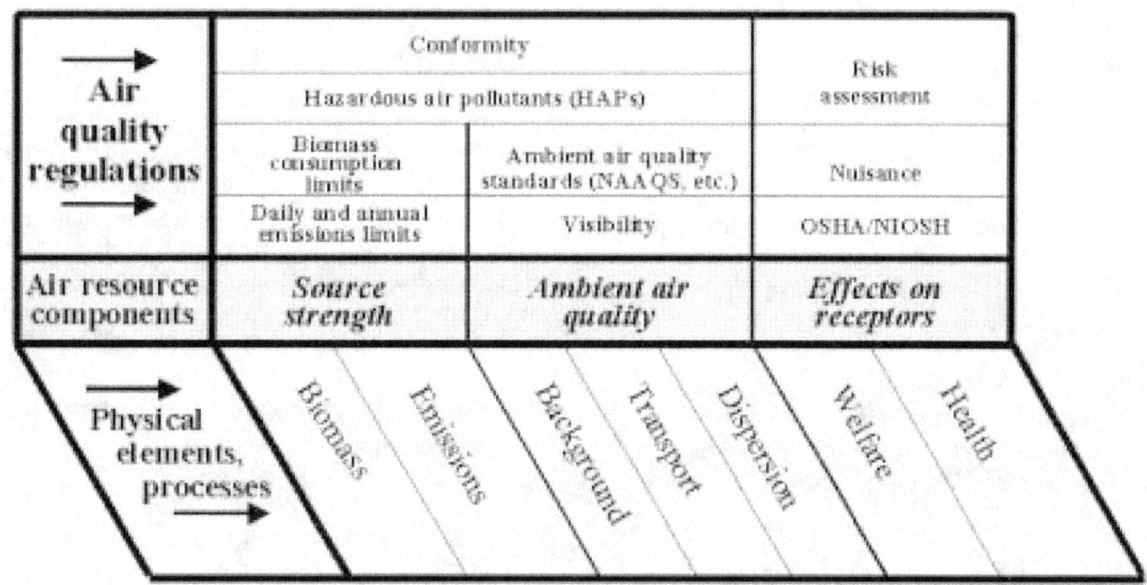

Figure 1-2—The relations of air regulations and physical processes to the three categories within the air resource component. OSHA/NIOSH = Occupational Safety and Health Administration/National Institute for Occupational Safety and Health (Sandberg and others 1999).

USDA Forest Service Gen. Tech. Rep. RMRS-GTR-42-vol. 5. 2002

3

sive treatment of the state of knowledge regarding fire and air quality and provides guidance to practitioners. We will not attempt to duplicate its level of detail in this volume. Rather, we add some technical background and analysis of research needs relative to new requirements for management.

First published in 1985, the guide is intended to provide national guidance for the planning and managing of smoke from prescribed fires to achieve air quality requirements through better smoke management practices (NWCG 1985). This guide has been widely distributed within the fire community and air quality regulatory agencies, and to private and Tribal land managers, providing a single comprehensive source of information on fire and air quality issues.

Much has changed since 1985 in prescribed burning practices, smoke management programs, and air quality regulatory requirements. These changes are reflected in the 2001 edition of the guide, which includes expanded sections on fire and emissions processes, smoke impacts on health, welfare, safety, and nuisance; regulations for smoke management; and the fundamentals of responsible smoke management (Hardy and others 2001). These fundamentals include fire planning, use of smoke management meteorology, techniques to reduce emissions, smoke dispersion prediction systems, air quality monitoring methods, and program assessment.

The most significant change in the guide is the expanded and updated section on techniques to reduce emissions and impacts. While the 1985 guide focused primarily on minimizing smoke impacts by meteorological scheduling and dispersion, the 2001 guide provides detailed information on emissions reduction techniques, used in different regions of the country, that have been useful, practicable, and effective in the field. This emphasis on actual reduction of emissions rather than dispersion was provided in response to air quality regulations that now target regional emissions reductions.

Readers will also find that the 2001 guide has a great deal more information on the latest developments in national air quality regulations that affect fire programs including the regional haze and visibility protection programs, Clean Air Act's conformity requirements, EPA's *Interim Air Quality Policy on Wildland and Prescribed Fires* (EPA 1998), and NEPA planning guidance. The guide was drafted by 16 authors and five editor/compilers working under the sponsorship of the NWCG Fire Use Working Team with support from the EPA.

Wildland Fire and Air Quality: National Strategic Plan

Another recent publication also provides a systematic review of the state of knowledge and information systems. This strategic plan was also sponsored by the NWCG and the Environmental Protection Agency (EPA).

In 1997, the NWCG Fire Use Working Team sanctioned a small group of fire research scientists and air quality managers to develop a *National Strategic Plan: Modeling and Data Systems for Wildland Fire and Air Quality* (Sandberg and others 1999) to foster development and implementation of models and data systems that could be used to manage air quality impacts of fires. The resulting report provides a conceptual design and strategic direction toward meeting the increasing need for information required to manage emissions from fire (Sandberg and others 1999). In November 1997, after 2 years of drafting and extensive review of a draft plan, 86 experts attended a national workshop, and using the discussion framework presented in this chapter, they defined the current state of knowledge, desired future condition, and recommendations for research and development for each cell in the discussion framework.

The strategic plan targets a more technical, scientific, and policy-oriented audience than the smoke management guide, and recommends a research and development strategy to reach a desired future state for smoke management information systems. It also provides a comprehensive treatment of policy and technical issues that we will not duplicate in this volume.

Introduction to Visibility

Air pollution impacts on visibility are discussed in detail in *Introduction to Visibility* (Malm 2000). The discussion is not specific to the impacts of fire but is relevant because of the regulatory attention given to fire in the EPA Regional Haze Rule (40 CFR Part 51 1999) and because Federal land managers have the responsibility of managing fires and the impacts of fires and all other pollution sources on visibility in many National Parks and wilderness areas. We make no attempt in this volume to duplicate this discussion of the atmospheric physics, meteorology, historic visibility trends, monitoring and apportionment methodologies, or human perceptions that are so admirably covered in *Introduction to Visibility*.

The Federal Advisory Committee Act White Papers

During the 1997 to 1998 development of proposed national ambient air quality standards (NAAQS) for PM2.5 (particulate matter with an aerodynamic diameter less than or equal to 2.5 microns) and regional haze regulations, EPA used provisions of the Federal Advisory Committee Act (FACA) to convene a large group of stakeholders who were interested in providing input to

4

USDA Forest Service Gen. Tech. Rep. RMRS-GTR-42-vol. 5. 2002

the regulatory process. A FACA committee for the development of ozone, particulate matter, and regional haze implementation programs was formed to address both policy and technical issues. The committee's Science and Technology Wildland Fire Issues Group, one of several working groups and subcommittees, researched and drafted five reports that are briefly summarized below (EPA 2000c).

Air Monitoring for Wildland Fire Operations provides recommendations for conducting air-monitoring programs designed to support fire activities that also monitor for compliance with NAAQS. It also describes how monitoring can support burning programs and how land managers can collaborate with air agencies, and it provides guidance for selecting monitoring equipment.

Elements of a Smoke Management Program discusses recommendations for a basic level smoke management program. The document summarized information from an EPA-sponsored workshop held to respond to specific questions posed by EPA. The document describes the six basic components of a smoke management program:

- Authorization to burn
- Minimizing emissions
- Burn plan components
- Public education
- Surveillance and enforcement
- Program evaluation

It also provides examples of monitoring methods, public awareness programs, and program enforcement.

Emission Inventories for State Implementation Plan [SIP] Development describes several levels of inventory complexity: a default level based on currently available information; a basic level program that is considered the minimal program needed to support SIP development; and a detailed inventory level when a greater level of analysis or accountability in inventory precision is needed. Elements of each level of inventory are described, data sources are identified and data management issues are discussed.

What Wildland Fire Conditions Minimize Emissions and Hazardous Air Pollutants and Can Land Management Goals Still be Met? This paper is a discussion of fire conditions and techniques that minimize pollutant emissions. Both wildland emissions and prescribed fire emissions are discussed. The discussion of emissions reduction techniques for prescribed burning is also found in *Smoke Management Guide for Prescribed and Wildland Fire: 2001 edition* (Hardy and others 2001).

Estimating Natural Emissions from Wildland and Prescribed Fire addresses how best to define "natural emissions" from fire. This is critical to implementing

regional haze goals of reducing visibility degradation caused by human-made sources of air pollution. The paper discusses a matrix of choices: (1) emissions from fire necessary to restore and sustain desired ecosystem characteristics, (2) fire needed to manage fuels to a condition where they can be dealt with most effectively from a wildfire control standpoint, (3) no net increase in fire emissions, and (4) no change from current emissions.

Stakeholders reviewed, discussed, and drafted, additional work on these five reports. The reports and other technical references were considered by EPA during the formulation of the regional haze regulations and revisions to the particulate matter NAAQS.

Environmental Regulation and Prescribed Fire Conference

In March 1995 a conference on new developments in environmental regulations related to prescribed fire was held in Tampa, FL (*Conference Proceedings: Environmental Regulation & Prescribed Fire: Legal and Social Challenges*, Bryan 1997). This 3-day meeting included sessions on challenges and strategies regarding the use of fire, air quality regulation, and liability, as well as social and economic issues. Sponsored by numerous State and Federal environmental and forestry agencies, the conference provided a forum for discussion of the Clean Air Act, Endangered Species Act, and other Federal statutes that guide national, State, and local regulations pertaining to prescribed fire.

Significantly, a joint declaration drafted by the conference steering committee and presented to conference attendees was later signed by representatives of the EPA, State of Florida, National Biological Survey, The Wilderness Society, Forest Service, and Mariposa County, Florida. In summary, the declaration upheld the following principles:

- Practitioner liability is a major obstacle to the increased use of fire. Legislation should be considered on the Federal level to protect properly certified fire practitioners except in cases where negligence is proven.
- Partnerships among all of the stakeholders are vital to the future use of fire. Efforts to enhance such partnerships must be encouraged especially in the exchange of information, development of best management practices, public education campaigns, and funding initiatives.
- Agencies should work together to evaluate tradeoffs between public health risks from fire and ecological damage caused by fire exclusion.

USDA Forest Service Gen. Tech. Rep. RMRS-GTR-42-vol. 5. 2002

5

- Public education regarding the use of prescribed fire, ecosystem health, and risks of wildfire versus those from prescribed burning is encouraged.
- The role of fire in ecosystem management needs to be understood by all stakeholders. The ramifications of not using prescribed fire are serious and must also be appreciated as limits on fire use may conflict with other public mandates.
- Actions pertaining to the use of fire must be based on sound science. There are several crucial knowledge gaps that must be filled. Consequences to public safety caused by delaying the increases of prescribed fire are great.
- Public and private property owners need to retain the right to use prescribed fire to protect and enhance the productivity of their lands while also protecting nearby property owners from adverse impacts of burning.
- Administrators responsible for allocating funds should do so on the basis of regional priorities with greater emphasis on prevention than in the past.
- An increased emphasis on training for prescribed fire practitioners is needed to enhance public acceptance.

Southern Forestry Smoke Management Guidebook

The *Southern Forestry Smoke Management Guidebook* (Southern Forest Fire Laboratory Personnel 1976) was one of the first smoke management guidebooks developed in the United States for use by land, fire, and air resources managers. The guide provides an improved understanding of: (1) smoke management and air quality regulations; (2) contents of smoke and variables affecting production; (3) smoke transport and dispersion; (4) potential effects on human health, human welfare, and visibility; and (5) what can be done to mitigate its impacts. A system for predicting and modifying smoke concentrations from prescribed fires was introduced for Southern fuels.

Changes in Fire Policy _____

The *Federal Wildland Fire Policy* (USDI and USDA 1995; USDI and others 2001) requires that "... fire, as a critical natural process, must be reintroduced into the ecosystem to restore and maintain sustainable ecosystems. This will be accomplished across agency boundaries and will be based on the best available science." The policy requires "the use of fire to sustain ecosystem health based on sound scientific principles and balanced with other social goals including public health and safety, air quality, and other specific environmental concerns." Early in the planning process, action is required to "involve public health and environmental regulators in developing the most workable application of policies and regulations." Agencies are called on to "create a system for coordination and cooperation among land managers and regulators that explores options within existing laws to allow for the use of fire to achieve goals of ecosystem health while protecting individual components of the environment, human health, and safety." The policy also requires that air quality values be considered during preparedness and fire protection. When setting protection priorities, land managers must "define values to be protected working in cooperation with state, local, and tribal governments, permittees, and public users. Criteria will include environmental, commodity, social, economic, political, public health, and other values."

Several strategies and funding programs were developed to improve the ability of managers to fully implement this policy.

Joint Fire Science Program

The Joint Fire Science Program (JFSP) was created by Congress in the 1998 Appropriations to Interior and Related Agencies bill to augment the delivery of science and information systems necessary to manage the increased use of fire and other fuel treatments. The legislation provides a mandate to protect air quality in conjunction with economic efficiency and ecological consequences. The program (National Interagency Fire Center 2002 unpaginated) recognizes that:

> Land managers are rapidly expanding the use of fire for managing ecosystems while air resource managers are accelerating efforts to reduce the local and regional impacts of smoke. Smoke management (meeting air quality standards) is a legal requirement of the Clean Air Act, as well as a health and safety issue for the general populace and fireline personnel. The JFSP will attempt to define these social relationships and develop analytical tools and communication practices to help mangers include social considerations in decision making.

One of the goals of the JFSP is "to evaluate various treatment techniques for cost effectiveness, ecological consequences, and air quality impacts." The program plan states:

> Methods have not been developed to assess the opportunities, costs, and effectiveness of employing smoke reduction techniques throughout the country. Current models to assess regional scale cumulative effects on air quality and water quality will need to be expanded. The program will develop a nationally consistent system of models for fuel consumption, emissions production, and smoke dispersal that can assess cumulative

6

USDA Forest Service Gen. Tech. Rep. RMRS-GTR-42-vol. 5. 2002

effect. This research would also contribute to understanding the potential national and global impacts of changes in biomass use, prescribed fire, and wildland fire on wood supply, atmospheric chemistry, and carbon sequestration.

Cohesive Strategy

Protecting People and Sustaining Resources in Fire-Adapted Ecosystems: A Cohesive Strategy (Laverty and Williams 2000) is the Federal framework established to restore and maintain ecosystem health to reduce the threat and consequences of wildfires. It is presumed that fire suppression over the past 100 years has excluded fire from many ecosystems, fueling conditions for unnaturally intense fires that, among other effects, threaten air quality. Citing serious air quality impacts from long duration wildfire episodes in recent years, the report expresses concern that:

> The extent to which management for ecosystem resilience can improve air quality over the long term is not completely known. Present regulatory policies measure prescribed fire emissions, but not wildland fire emissions. The emissions policy tends to constrain treatments and – in short interval fire systems — may act to inadvertently compound wildland fire risks. (p 34)

The cohesive strategy directs land management agencies to collaborate with the EPA in addressing long-term impacts, tradeoffs, and issues regarding air quality and other impacts. The report acknowledges that programmatic analysis of air quality impacts will be a necessary step in implementing the planned increases in prescribed burning necessary to restore the health of fire-prone ecosystems. The strategy estimates that the USDA Forest Service Regions would increase fuel treatments by five-fold in the West and two-fold in the East and South to achieve restoration goals within 10 years; or employ a slightly smaller increase to obtain results in 20 years. Most, but not all, of the treatments would involve burning.

The relative risk to air quality was projected to decrease by about 25 percent as a result of improving the resilience of ecosystems, according to current models.

The cohesive strategy is responsive to regulatory responsibilities. The planned increase in burning is constrained in part by the consideration to regulatory obligations, with an acknowledgment that a more rigorous assessment of impacts could substantially change the planned extent and schedule of treatments. Concerns for public health issues and firefighter safety in relation to smoke are also expressed. The strategy acknowledges that air quality issues must be analyzed more thoroughly at smaller scales as it is stepped down to landscape and project level planning.

National Fire Plan

The National Fire Plan was established in *A Report to the President In Response to the Wildfires of 2000* (USDA and USDI 2000), and implemented using *Collaborative Approach for Reducing Wildfire Risks to Communities and the Environment: 10-Year Comprehensive Strategy* (Western Governors' Association 2001). Stakeholder groups under the sponsorship of the USDA Forest Service, USDI, and the Western Governors' Association prepared the implementation strategy. This strategy recognizes that key decisions in setting priorities for restoration, fire, and fuel management should be made at local levels. As such, the plan requires an ongoing process whereby the local, Tribal, State and Federal land management, scientific, and regulatory agencies exchange the required technical information, including the assessment of air quality tradeoffs, to inform this decisionmaking process. The strategy has a goal of maintaining and enhancing community health and economic and social well-being; and requires that public health risks from smoke are reduced, airshed visibility is improved, and smoke management plans are developed in conjunction with prescribed fire planning and implementation.

USDA Forest Service Gen. Tech. Rep. RMRS-GTR-42-vol. 5. 2002

7

Notes

_____

_____

_____

_____

_____

_____

Chapter 2: Air Quality Regulations and Fire

This chapter introduces the regulatory environment for smoke from prescribed and wildland fire, providing updated discussion of the laws, regulations, standards, and regulatory strategies that have changed since about 1980. We explain roles and responsibilities of the regulatory agencies and land managers, and we frame the technical discussion in the context of who needs what information to fulfill legal requirements.

Air pollution is the presence in the atmosphere of one or more contaminants of a nature, concentration, and duration to be hazardous to human health or welfare (Sandberg and others 1999). Welfare includes potential to harm animal or ecosystem health, economic activity, or the comfortable enjoyment of life and property. Air pollution is created from both human (that is, anthropogenic) and natural sources. Anthropogenic air pollution is the presence in the atmosphere of a substance or substances added directly or indirectly by a human act, in such amounts as to adversely affect humans, animals, vegetation, or materials (Williamson 1973). Air pollutants are classified into two major categories: primary and secondary. Air pollutant emissions, or simply "emissions," are the production and release of air contaminants emitted from fires that have a potential to cause air pollution. This definition includes particulates, hydrocarbons, carbon monoxide (CO), metals, and all other

trace gases that may be hazardous or that are chemical precursors to secondary air pollution. Primary pollutants are those directly emitted into the air. Under certain conditions, primary pollutants undergo chemical reactions within the atmosphere and produce new substances known as secondary pollutants. Hazardous air pollutants are a special class of air pollutants identified in the Clean Air Act Amendments of 1990 as constituting a hazard to human health.

Air quality is a measure of the presence of air pollution. Ambient air quality is defined by the Clean Air Act of 1963 as the air quality anywhere people have access, outside of industrial site boundaries. Ambient air quality standards are standards of air quality designed to protect human health or welfare. Air resource management includes any activity to anticipate, regulate, or monitor air pollution, air pollutant emissions, ambient air quality, or the effects of air pollution resulting from fires or fire management.

In the past, emissions from prescribed fire were considered human-caused, and wildland fires were considered natural sources of emissions. But recent policy debate has focused on what should be considered natural; that is, to be reasonably unaffected by human influence. This debate resulted from the paradox that not all wildland fires are vigorously

USDA Forest Service Gen. Tech. Rep. RMRS-GTR-42-vol. 5. 2002

9

suppressed and that some prescribed burning is done to maintain healthy natural ecosystems where fire has previously been excluded.

Air resource management includes any activity to anticipate, regulate, or monitor air pollution, air pollutant emissions, ambient air quality, or the effects of air pollution resulting from fires or fire management.

Emissions and impacts on air quality from fires are managed and regulated through a complex web of interrelated laws and regulations. The primary legal basis for air quality regulation across the nation is the Federal Clean Air Act (CAA), which is actually a series of acts, amendments, and regulations that include:

- Federal Air Pollution Control Act of 1955 (PL 84-159). Provides for research and technical assistance and authorizes the Secretary of Health, Education, and Welfare to work toward a better understanding of the causes and effects of air pollution.
- Federal Clean Air Act of 1963 (PL 88-206). Empowers the Secretary of Health, Education, and Welfare to define air quality criteria based on scientific studies. Provides grants to state and local air pollution control agencies.
- Federal Air Quality Act of 1967 (PL 90-148). Establishes a framework for defining "air quality control regions" based on meteorological and topographical factors of air pollution.
- Federal Clean Air Act Amendments of 1970 (PL 91-604). Principal source of statutory authority for controlling air pollution. Establishes basic U.S. program for controlling air pollution.
- Environmental Protection Agency (EPA) promulgates national ambient air quality standards (NAAQS) for particulates, photochemical oxidants (including ozone), hydrocarbons, carbon monoxide, nitrogen dioxide, and sulfur dioxide (1971).
- Clean Air Act Amendments of 1977 (PL 95-95). Sets the goal for visibility protection and improvement in Class I areas and assigns Federal land managers the affirmative responsibility to protect air quality related values.
- Clean Air Act Amendments of 1990 (PL 101-549). Establishes authority for regulating regional haze and acknowledges the complexity of the relation between prescribed and wildland fires.
- Regional Haze Regulations, Final Rule (40 CFR Part 51) (1999). EPA promulgates the Regional Haze Rule supported in part by the 1998 Interim Air Quality Policy on Wildland and Prescribed Fires.

Roles and Responsibilities Under the Clean Air Act

States have the lead in carrying out provisions of the Clean Air Act because appropriate and effective design of pollution control programs requires an understanding of local industries, geography, transportation, meteorology, urban and industrial development patterns, and priorities. The EPA has the task of setting air quality standards (national ambient air quality standards, or NAAQS). In addition, EPA develops policy and technical guidance describing how various Clean Air Act programs should function and what they should accomplish. States develop State implementation plans (SIPs) that define and describe customized programs they will implement to meet requirements of the Clean Air Act. Tribal lands are legally equivalent to State lands, and Tribes prepare Tribal implementation plans (TIPs) to describe how they will implement the Clean Air Act. Individual States and Tribes can require more stringent air quality standards but cannot weaken clean air goals set by EPA.

Federal land managers have the complex role of managing a fire as a source of air pollutants, while fulfilling monitoring and regulatory responsibilities tied to visibility and regional haze. Federal land managers are given the responsibility by the Clean Air Act for reviewing prevention of significant deterioration (PSD) permits (discussed later in this chapter) of major new and modified stationary pollution sources and commenting to the State on whether there is concern for visibility impacts (or other resource values) in Class I areas downwind of the proposed pollution source. Some States require modeling of source impacts on Class I areas, and Federal land managers customarily comment on the model results.

The 1990 Clean Air Act Amendments require planned Federal actions to conform to SIPs. This "general conformity rule" prohibits Federal agencies from taking any action within a nonattainment or maintenance area that (1) causes or contributes to a new violation of air quality standards, (2) increases the frequency or severity of an existing violation, or (3) delays the timely attainment of a standard as defined in the applicable SIP or area plan. The general conformity rule covers direct and indirect emissions of criteria pollutants, or their precursors, which are caused by a Federal action, are reasonably foreseeable, and can practically be controlled by the Federal agency through its continuing program responsibility.

10

USDA Forest Service Gen. Tech. Rep. RMRS-GTR-42-vol. 5. 2002

National Ambient Air Quality Standards

The purpose of the Clean Air Act is to protect humans against negative health or welfare effects from air pollution. National ambient air quality standards (NAAQS) are defined in the Clean Air Act as amounts of pollutant above which detrimental effects to public health or welfare may result. NAAQS have been established for the following criteria pollutants: particulate matter (PM10 and PM2.5; NAAQS for particulate matter are established for two aerodynamic diameter classes: PM10 is particulate matter less than 10 microns in diameter, and PM2.5 is less than 2.5 microns in diameter; total suspended particulate matter is called PM or sometimes TSP), sulfur dioxide (SO_2), nitrogen dioxide (NO_2), ozone, carbon monoxide (CO) and lead (Pb) (table 2-1). Primary NAAQS are set at levels to protect human health; secondary NAAQS are to protect human welfare effects including visibility as well as plant and materials damage.

An area that is found to be in violation of a primary NAAQS is labeled a nonattainment area (fig. 2-1); an area once in nonattainment but recently meeting NAAQS, and with appropriate planning documents approved by EPA, is a maintenance area; all other areas are attainment or unclassified (due to lack of monitoring). State air quality agencies can provide up-to-date locations of local nonattainment areas (PM2.5 is a newly regulated pollutant, so attainment/nonattainment status had not been determined at the time of publication of this document; monitoring must take place for at least 3 years before designation can be made, which means PM2.5 status will likely not be known until at least 2003). States are required through their SIPs to define programs for implementation, maintenance, and enforcement of the NAAQS within their boundaries. Wildland fire in and near nonattainment areas will be scrutinized to a greater degree than in attainment areas and may be subject to general conformity rules. Extra planning, documentation, and careful scheduling of prescribed fires will likely be required to minimize smoke effects in the nonattainment area to the greatest extent possible. In some cases, the use of fire may not be possible if significant impacts to a nonattainment area are likely.

The major pollutant of concern in smoke from fire is fine particulate matter, both PM10 and PM2.5. Studies indicate that 90 percent of all smoke particles emitted during wildland burning are PM10, and 90 percent of PM10 is PM2.5 (Ward and Hardy 1991). The most recent human health studies on the effects of particulate matter indicate that fine particles, especially PM2.5, are largely responsible for health effects including mortality, exacerbation of chronic disease, and increased hospital admissions (Dockery and others 1993; Schwartz and others 1996).

Prevention of Significant Deterioration

Another provision of the Clean Air Act with some applicability to wildland burning activities is the prevention of significant deterioration (PSD) provisions.

Table 2-1—National ambient air quality standards (NAAQS) (U.S. Environmental Protection Agency 2000b). Primary NAAQS are set at levels to protect human health; secondary NAAQS are to protect human welfare.

Pollutant	Averaging time	Primary	Secondary
PM10	Annual arithmetic mean	50 $\mu g/m^3$ [a]	50 $\mu g/m^3$
	24-hour average	150 $\mu g/m^3$	150 $\mu g/m^3$
PM2.5	Annual arithmetic mean	15 $\mu g/m^3$	15 $\mu g/m^3$
	24-hour average	65 $\mu g/m^3$	65 $\mu g/m^3$
Sulfur dioxide (SO_2)	Annual average	0.03 ppm [b]	—
	24-hour average	0.14 ppm	—
	3-hour average	—	0.50 ppm
Carbon monoxide (CO)	8-hour average	9 ppm	—
	1-hour average	35 ppm	—
Ozone (O_3)	8-hour average	0.12 ppm	0.12 ppm
	1-hour average	0.08 ppm	0.08 ppm
Nitrogen dioxide (NO_2)	Annual average	0.053 ppm	0.053 ppm
Lead (Pb)	Quarterly average	1.5 $\mu g/m^3$	1.5 $\mu g/m^3$

[a] $\mu g/m^3$ = micrograms per cubic meter.
[b] ppm = parts per million.

USDA Forest Service Gen. Tech. Rep. RMRS-GTR-42-vol. 5. 2002

11

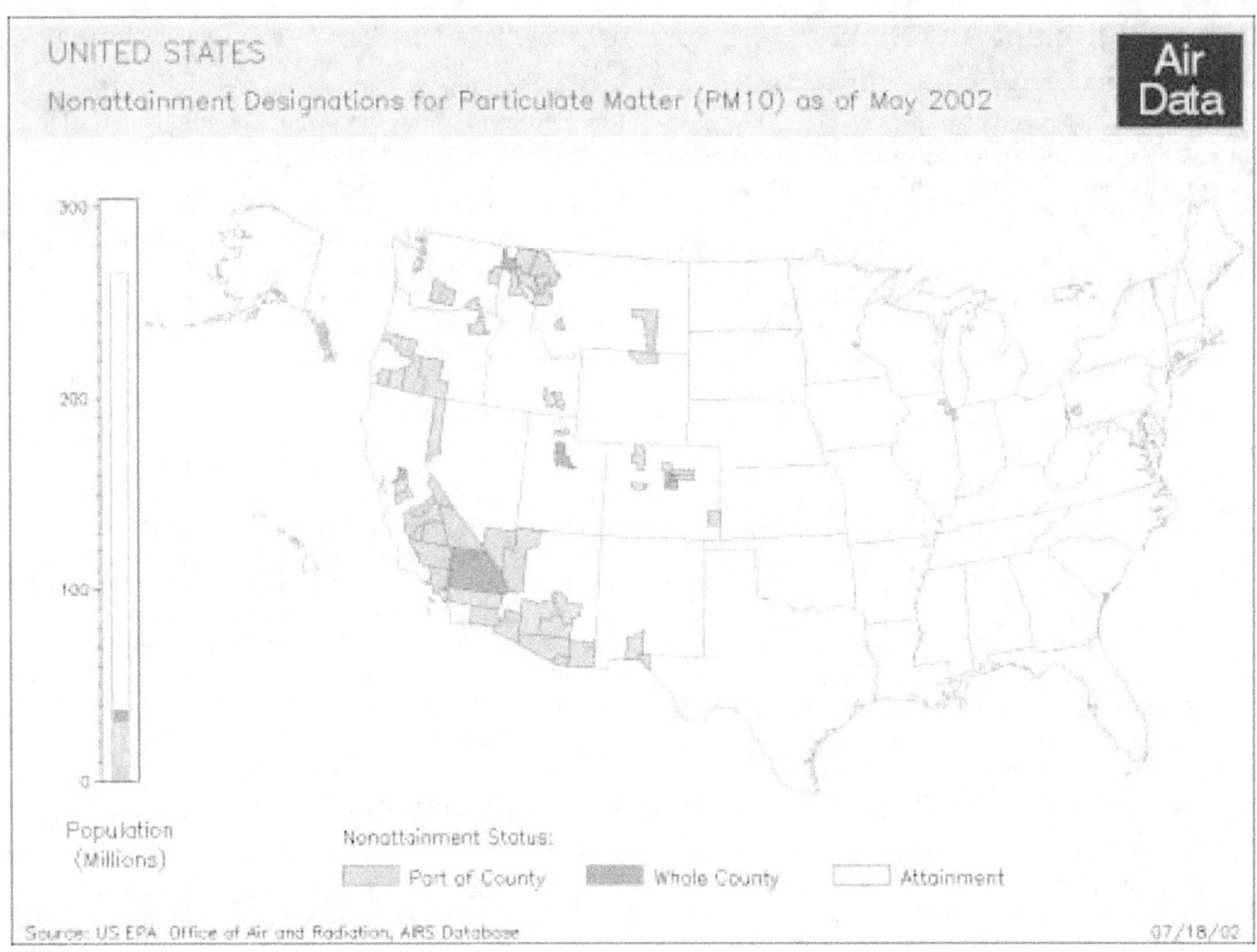

Population
(Millions)

Nonattainment Status:

Part of County   Whole County   Attainment

Source: US EPA Office of Air and Radiation, AIRS Database   07/18/02

Figure 2-1—PM10 nonattainment areas as of May 2002. Current nonattainment status for PM10 and all other criteria pollutants are available from the Environmental Protection Agency (EPA) aerometric information retrieval system (AIRS) Web page at http://www.epa.gov/air/data/index.html (EPA 2002).

The goal of PSD is to prevent areas that are currently cleaner than is allowed by the NAAQS from being polluted up to the maximum ceiling established by the NAAQS. Three air quality classes were established by the Clean Air Act PSD provisions including Class I (which allows very little additional pollution), Class II (which allows some incremental increase in pollution), and Class III (which allows pollution to increase up to the NAAQS). Class I areas include wildernesses and national memorial parks over 5,000 acres, National Parks exceeding 6,000 acres, and all international parks that were in existence on August 7, 1977, as well as later expansions to these areas (fig. 2-2).

Historically, EPA has regarded smoke from wildland fires as temporary and therefore not subject to issuance of a PSD permit; whether or not wildland fire smoke should be considered when calculating PSD increment consumption or PSD baseline was not defined. EPA recently reaffirmed that States could exclude prescribed fire emissions from increment analyses provided the exclusion does not result in permanent or long-term air quality deterioration (EPA 1998). States are also expected to consider the extent to which a particular type of burning activity is truly temporary, as opposed to an activity that could be expected to occur in a particular area with some regularity over a long period. Oregon is the only State that has chosen to include prescribed fire emissions in PSD increment and baseline calculations.

Visibility

The 1977 amendments to the Clean Air Act include a national goal of "the prevention of any future, and the remedying of any existing, impairment of visibility

12

USDA Forest Service Gen. Tech. Rep. RMRS-GTR-42-vol. 5. 2002

NP	National Park
IP	International Park
NM	National Monument
NS	National Seashore
W	Wilderness

Figure 2-2—Mandatory class 1 areas (Hardy and others 2001).

in mandatory Class I Federal areas which impairment results from manmade air pollution" (42 U.S.C § 7491). States are required to develop implementation plans that make "reasonable progress" toward the national visibility goal.

Atmospheric visibility is affected by scattering and absorption of light by particles and gases. Particles and gases in the air can obscure the clarity, color, texture, and form of what we see. Fine particles most responsible for visibility impairment are sulfates, nitrates, organic compounds, elemental carbon (or soot), and soil dust. Sulfates, nitrates, organic carbon, and soil tend to scatter light, whereas elemental carbon tends to absorb light. Fine particles (PM2.5) are more efficient per unit mass than coarse particles (PM10 and larger) at causing visibility impairment. Naturally occurring visual range in the Eastern United States is estimated to be between 60 and 80 miles, while natural visual range in the Western United States is between 110 and 115 miles (these estimates do not consider the effect of natural fire on visibility)

(Trijonis and others 1991). Currently, visual range in the Eastern United States is about 15 to 30 miles and about 60 to 90 miles in the Western United States. (40 CFR Part 51). The theoretical maximum visual range about 240 miles.

Regional Haze

Regional haze is visibility impairment produced by a multitude of sources and activities that emit fine particles and their precursors and are located across a broad geographic area. This contrasts with visibility impairment that can be traced largely to a single, large pollution source. Until recently, the only regulations for visibility protection addressed impairment that is reasonably attributable to a permanent, large emissions source or small group of large sources. In 1999, EPA issued regional haze regulations to manage and mitigate visibility impairment from the multitude of diverse regional haze sources (40 CFR Part 51). The regional haze regulations call for States to establish

USDA Forest Service Gen. Tech. Rep. RMRS-GTR-42-vol. 5. 2002

13

goals for improving visibility in Class I National Parks and wildernesses, and to develop long-term strategies for reducing emissions of air pollutants that cause visibility impairment.

Regional Haze Planning Process—Because regional haze is a multi-State issue, regional haze regulations encourage States, land managers, and other stakeholders to work together to develop control programs through regional planning organizations that can coordinate development of strategies across a multi-State region. In the Western United States, the Western Regional Air Partnership (WRAP), sponsored through the Western Governors' Association and the National Tribal Environmental Council, is coordinating regional planning and technical assessments. The WRAP was the first of five regional planning organizations to be established and has been active in many technical and policy developments. Other regional planning organizations have begun assessments of fire and air quality in their regions. In the Eastern United States, four formal groups are addressing planning issues: CENRAP (Central States Regional Air Partnership), OTC (Ozone Transport Commission), VISTAS (Visibility Improvement State and Tribal Association of the Southeast); and the Midwest Regional Planning Organization (fig. 2-3).

As inter-State smoke transport becomes a larger issue, agencies are expanding coordination of their burns. Multi-State, interagency partnerships are developing to help coordinate burning and mitigate cumulative impacts of smoke. For example, the Montana/Idaho airshed group includes private, State, Tribal, and Federal partners in supporting an integrated smoke management program that includes emissions monitoring and smoke forecasting (Levinson 2001).

Regional Haze and Fire Emissions—The adoption of regional haze regulations marks a turning point in how fire emissions are treated under the nation's Federal and State air quality regulations,

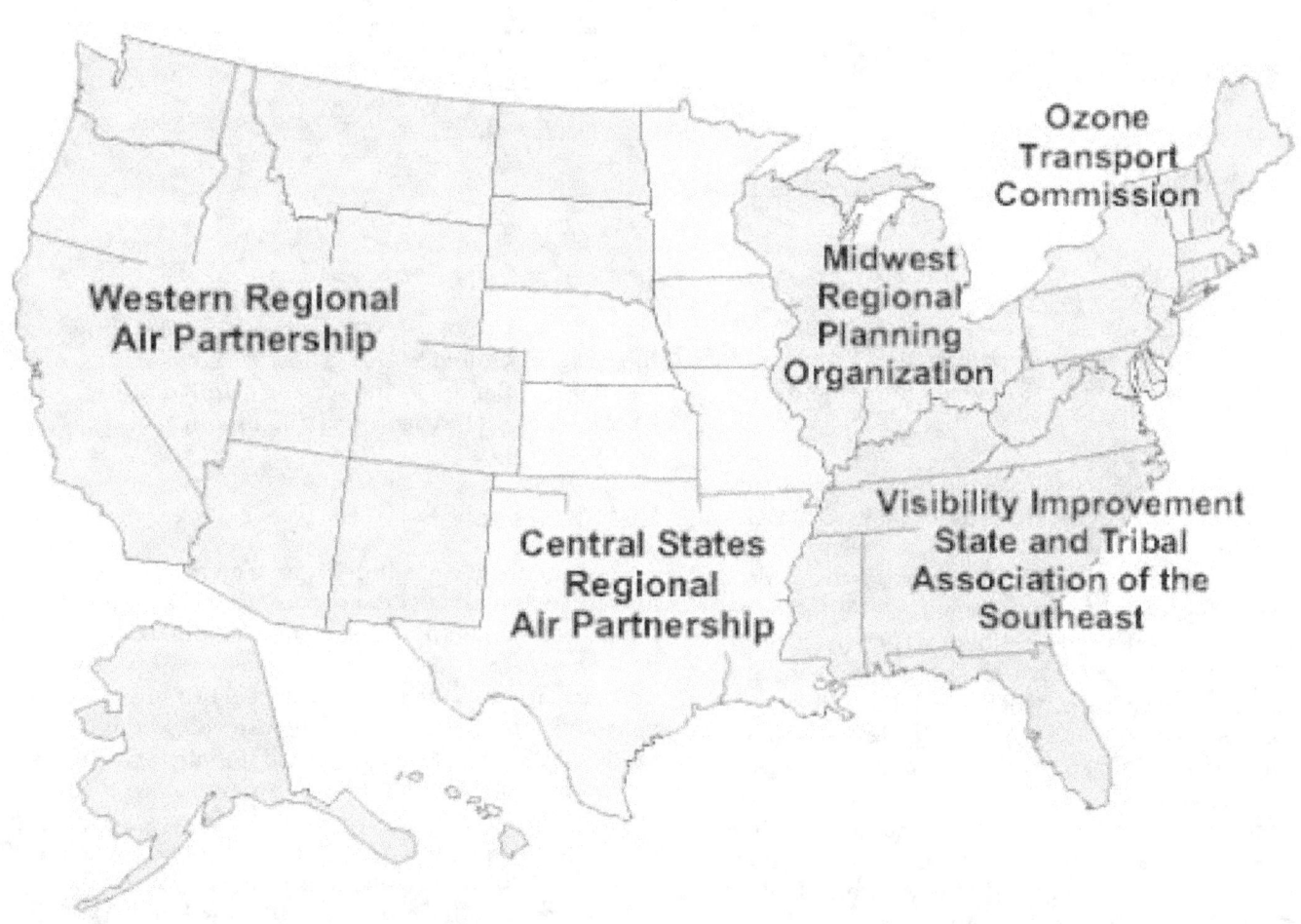

Figure 2-3—Regional air quality planning groups (Hardy and others 2001).

14

USDA Forest Service Gen. Tech. Rep. RMRS-GTR-42-vol. 5. 2002

although the regulations leave several definitions open to subsequent policy interpretation:

- The role of fire in forest ecosystems is formally recognized for the first time.
- Emissions from "natural" sources are distinguished from "anthropogenic" sources and treated differently under the rule.
- The rule is the first to require development of emissions inventories for fire, including wildland fires.
- Emissions from fire are now subject to regional air quality planning processes as well as requirements to achieve "reasonable progress" in emissions reductions

The policy discussion to determine what types of fire emissions are considered natural is still in progress, but the WRAP has recommended a national policy that would (1) define "natural background" as fire emissions that would occur in the future without fire management; that is, without reference to historic fire occurrence or historic vegetation types; and (2) include prescribed burning as natural sources of visibility impacts when fire is used to maintain healthy and sustainable ecosystems.

Current data from a national visibility-monitoring network (Sisler and others 1996) do not show fire to be the predominant long-term source of visibility impairment in any Class I area (40 CFR Part 51), although emissions from fire are an important episodic contributor to visibility-impairing aerosols. Certainly the contribution to visibility impairment from fires can be significant over short periods, but fires in general occur relatively infrequently and thus have a lesser contribution to long-term averages. Specific goals for visibility improvement focus efforts on improving air quality on the most impaired days, so fires may prove to be an important target for control efforts in some areas

Fire Consortia for Advanced Modeling of Meteorology and Smoke (FCAMMS)—Multiagency consortia are building in the Pacific Northwest, Rocky Mountain region, and Northeastern and Southeastern United States as part of the U.S. Department of Agriculture, Forest Service, Fire Consortia for Advanced Modeling of Meteorology and Smoke. The Pacific Northwest consortium is developing a real-time smoke prediction and emission tracking system that addresses needs of several smoke management plans from collaborating States, Tribes, and local air agencies (Ferguson and others 2001). California and Nevada are working together through the California and Nevada Smoke and Air Committee (CANSAC) with similar objectives of tracking and predicting cumulative smoke impacts (Chris Fontana, personal communication).

Each group or regional consortium must respond to local, State, and Tribal smoke management programs. In addition, each region of the country has its own particular atmospheric processes that impact fire behavior and smoke dispersion in different ways. For example, while in the Southeast, timing of frontal passages and onshore flow regimes become critical, in the Western United States, complex flow through mountainous terrain is an important consideration in managing smoke. These regionally specific demands are forcing research to focus on subtle aspects of smoke emissions and dispersion instead of traditional development of worst-case air pollution scenarios.

Reasonable Progress

Visibility rules require States to make "reasonable progress" toward the Clean Air Act goal of "prevention of any future, and the remedying of any existing, impairment of visibility." The regional haze regulations did not define visibility targets but instead gave States flexibility in determining reasonable progress goals for Class I areas. States are required to conduct analyses to ensure that they consider the possibility of setting an ambitious reasonable progress goal, one that is aimed at reaching natural background conditions in 60 years. The rule requires States to establish goals for each affected Class I area to (1) improve visibility on the haziest 20 percent of days, and (2) ensure no degradation occurs on the clearest 20 percent of days over the period of each implementation plan.

States are to analyze and determine the rate of progress needed for the implementation period extending to 2018 such that, if maintained, this rate would attain natural visibility conditions by the year 2064. To calculate this rate of progress, each State must compare baseline visibility conditions to estimate natural visibility conditions in Class I areas and to determine the uniform rate of visibility improvement that would need to be maintained during each implementation period to attain natural visibility conditions by 2064. Baseline visibility conditions will be determined from data collected from a national network of visibility monitors representing all Class I areas in the country for the years 2000 to 2004. Each State must determine whether this rate and associated emissions reduction strategies are reasonable based on several statutory factors. If the State finds that this rate is not reasonable, it must provide a demonstration supporting an alternative rate.

Hazardous Air Pollutants _____

Hazardous air pollutants (HAPs) are identified in Title III of the Clean Air Act Amendments of 1990

USDA Forest Service Gen. Tech. Rep. RMRS-GTR-42-vol. 5. 2002

15

(PL 101-549) as 188 different pollutants "which present, or may present, through inhalation or other routes of exposure, a threat of adverse human health or environmental effects whether through ambient concentrations, bioaccumulation, deposition, or other routes." The list of HAPs identified in the Clean Air Act are substances that are known or suspected to be carcinogenic, mutagenic, teratogenic, neurotoxic, or which cause reproductive dysfunction.

EPA Interim Air Quality Policy on Wildland and Prescribed Fires

In 1998, the EPA issued a national policy to address how best to achieve national clean air goals while improving the quality of wildland ecosystems through the increased use of fire. The *Interim Air Quality Policy on Wildland and Prescribed Fires* (U.S. Environmental Protection Agency 1998) was developed through a partnership effort involving EPA, the U.S. Departments of Agriculture, Defense, and the Interior, State foresters, State and Tribal air regulators, and others. The group that developed the policy relied on the assumption that properly managed prescribed fires can improve the health of wildland ecosystems and reduce the health and safety risks associated with wildfire, while meeting clean air and public health goals through careful planning and cooperation among land managers, air quality regulators, and local communities.

Natural Events Policy

PM10 NAAQS exceedances caused by natural events are not counted toward nonattainment designation if a State can document that the exceedance was truly caused by a natural event and prepares a natural events action plan (NEAP) to address human health concerns during future events (Nichols 1996). Natural events are defined by this policy as wildfire, volcanic, seismic, and high wind events.

A wildfire NEAP should include commitments by the State and stakeholders to:

1. Establish public notification and education programs.
2. Minimize public exposure to high concentrations of PM10 due to future natural events such as by:
   a. Identifying the people most at risk.
   b. Notifying the at-risk public that an event is active or imminent.
   c. Recommending actions to be taken by the public to minimize their pollutant exposure.
   d. Suggesting precautions to take if exposure cannot be avoided.

3. Abate or minimize controllable sources of PM10 including the following:
   a. Prohibition of other burning during pollution episodes caused by wildfire.
   b. Proactive efforts to minimize fuel loadings in areas vulnerable to fire.
   c. Planning for prevention of NAAQS exceedances in fire management plans.
4. Identify, study, and implement practical mitigating measures as necessary.
5. Periodic reevaluation of the NEAP.

Collaboration Among Stakeholders

Because smoke from fire can negatively affect public health and welfare, air quality protection regulations must be understood and followed by responsible fire managers. Likewise, air quality regulators need an understanding of how and when fire use decisions are made and should become involved in fire and smoke management planning processes, including the assessment of when and how alternatives to fire will be used. Cooperation and collaboration between fire managers and air quality regulators is of great importance. Table 2-2 contains recommendations for various types of cooperation by these two groups depending on the applicable air quality protection instrument.

Best Available Control Measures

The application of best available control measures (BACM) for prescribed fire is a required element of State implementation plans for PM10 nonattainment areas that are significantly impacted by prescribed fire smoke (EPA 1992a). The application of BACM is also a requirement of EPA's *Air Quality Policy on Wildland and Prescribed Fires* (EPA 1998) (see "Prior Work" section in chapter 1). EPA's BACM guidance includes basic smoke management program elements and emissions reduction techniques that can be used by land managers to minimize air quality impacts from fire. These program elements and emissions reduction techniques are fully documented in the *Smoke Management Guide for Prescribed and Wildland Fire: 2001 Edition* (Hardy and others 2001).

Briefly, the BACM guidance notes that there are two basic approaches to minimizing the impact of prescribed fire on air quality: reducing the amount of pollutants emitted, or reducing the impact of the pollutants emitted on sensitive locations or regional haze through smoke dilution or transport (redistributing emissions). Although each method can be discussed

16

USDA Forest Service Gen. Tech. Rep. RMRS-GTR-42-vol. 5. 2002

Table 2-2—Recommended cooperation between wildland fire managers and air quality regulators, depending on air quality protection instrument (Hardy and others 2001).

Air quality protection instrument	Wildland fire managers	Air quality regulators
National ambient air quality standards (NAAQS)	Aware[a]	Lead[b]
Attainment status	Aware	Lead
State implementation plan (SIP) planning and development	Involved[c]	Lead
Conformity	Involved	Lead
Smoke management programs	Partner[d]	Lead
Visibility protection	Involved	Lead
Regional planning groups	Partner	Lead
Natural emissions	Partner	Lead
Natural events action plan	Partner	Lead
Land use planning	Lead	Involved
Project NEPA documents	Lead	Involved
Other fire planning efforts	Lead	Involved

[a]Aware: Responsibility to have a complete working knowledge of the air quality protection instrument but likely little or no involvement in its development or daily implementation.

[b]Lead: Responsibility to initiate, bring together participants, complete, and implement the particular air quality protection instrument.

[c]Involved: Responsibility to participate in certain components of development and implementation of the air quality protection instrument although not at full partner status.

[d]Partner: Responsibility to fully participate with lead organization toward development and implementation of the air quality protection instrument in a nearly equal relationship.

independently, fire practitioners often choose fire and fuels manipulation techniques that complement or are at least consistent with meteorological scheduling for maximum smoke dispersion and favorable plume transport. The following emissions reduction and redistributing emissions techniques are a compilation of our knowledge base, and depending on specific fire use objectives, the project locations, time, and cost constraints may or may not be applicable.

Reducing Emissions

At least 24 methods within six major classifications have been used to reduce emissions from prescribed burning (Hardy and others 2001). These techniques include methods designed to minimize emissions by reducing the area burned; reducing the fuel load by reducing the fuel production, or fuel consumption, or both; scheduling burns before new fuels appear; and increasing combustion efficiency. Each of these methods has specific practices associated with it.

Redistributing Emissions

These measures are commonly practiced in smoke management programs and include burning when dispersion is good, cooperating with other burners in

a single airshed to schedule burns, avoiding sensitive areas, burning smaller units, and burning more frequently.

Ozone and Fire

Ozone is a criteria air pollutant, but there is little monitoring or research data that directly link fire emissions with ground-level ozone concentrations. Regulating efforts to reduce ozone have therefore focused on more obvious industrial and urban sources of the pollutants that form ozone (NO_X and VOCs). Fires are known to emit VOCs and a minor amount of NO_X, but much is uncertain about the magnitude of ozone formation in the plume, the degree of mixing with urban sources of ozone precursors, and transport of ozone to ground level. EPA plans to begin including fire emissions in future regional ozone strategy modeling. Field observations of ozone formation in smoke plumes from fires date back nearly 25 years when measurements from aircraft detected ozone at the edge of forest fire smoke plumes aloft. A recent study (Wotawa and Trainer 2000) did link high ground-level ozone concentrations to forest fire plumes that had been transported great distances. Chapter 6 explores these issues more fully.

USDA Forest Service Gen. Tech. Rep. RMRS-GTR-42-vol. 5. 2002

17

Notes

_____

_____

_____

_____

_____

_____

18

USDA Forest Service Gen. Tech. Rep. RMRS-GTR-42-vol. 5. 2002

Chapter 3: Overview of Air Pollution from Fire

This chapter provides a brief overview of and an appreciation for the national, regional, and local importance of smoke to ambient air quality. We discuss the significance of fire emissions and air quality impacts on a national and regional scale. Chapter 7 of this document adds additional depth to this discussion.

Magnitude of Fire Contributions_____

Air quality impacts associated with wildland fires are distinguished from those resulting from prescribed burning because emissions from these two sources have in the past been treated differently under the Clean Air Act and by State and local air quality regulations. In addition, it is important to have a historical perspective of these issues given the increased use of fire in the recent past.

A comparison by Leenhouts (1998) of estimated levels of biomass burning suggests that 10 times more area burned annually in the pre industrial era than in the contemporary era. After accounting for land use changes such as urbanization and agriculture, Leenhouts concluded that about 50 percent of historical levels would burn today if historical fire regimes were restored to all wildlands to maintain ecosystem health (figs. 3-1 and 3-2). This suggests a four- to six-fold increase from the current magnitude of wildland fire emissions.

This section discusses: (1) smoke from wildland fires; (2) smoke from prescribed fires; (3) impacts on national ambient air quality standards (NAAQS); (4) and magnitude with respect to regional and subregional scale visibility degradation. The second section discusses smoke management programs.

Smoke from Wildland Fires

Although wildland fires occur throughout the nation, the largest fires and greatest number of fires occur in Alaska, the Southeastern States, and the West. Figure 3-3 shows the location of major fires during the 2000 fire season when 90,674 fires burned 7,259,159 acres (2,938,931 ha) at a fire suppression cost of $1.6 billion. The 10-year average acreage burned between 1990 and 1999 was 3.78 million acres (1.53 million ha), testifying to the severity of the 2000 wildfire season. Figure 3-4 shows those States that had more than 100,000 acres (40,486 ha) burned per year, on average, over the 1987 through 1997 period, illustrating that Alaska wildfires burn far more acres than fires in any other State. Area burned in California, the States in the Intermountain West, Florida, and the Southwest follow (Peterson 2000).

USDA Forest Service Gen. Tech. Rep. RMRS-GTR-42-vol. 5. 2002

19

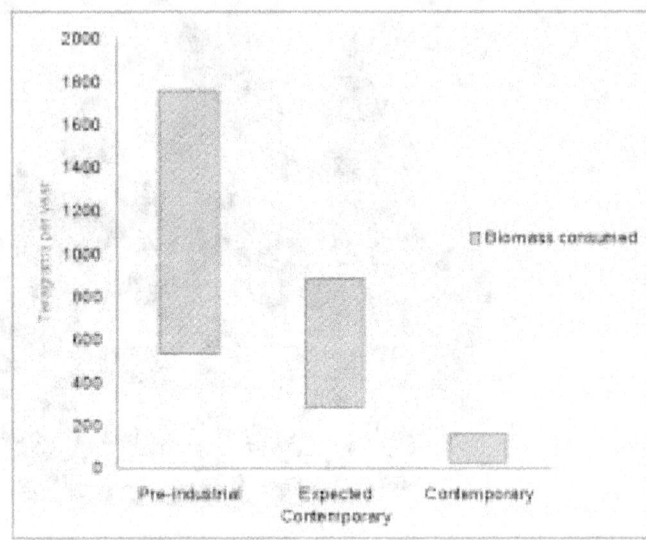

Figure 3-1—Estimated annual preindustrial, expected contemporary, and contemporary area (Mha) for the conterminous United States (from Leenhouts 1998).

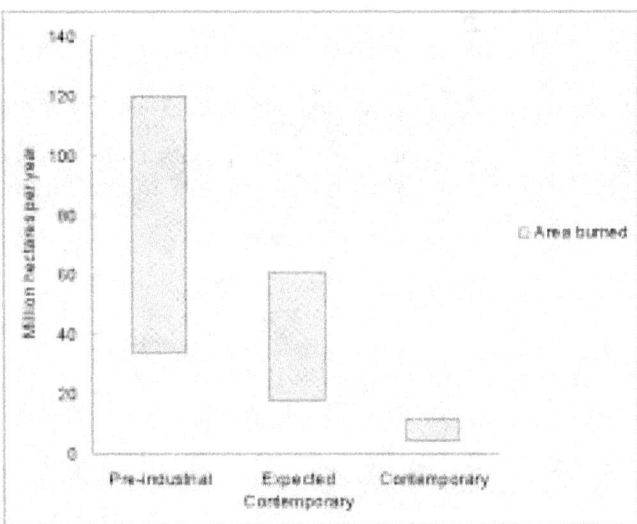

Figure 3-2—Estimated annual preindustrial, expected contemporary, and contemporary biomass consumed (Tg x 10^2) for the conterminous United States (from Leenhouts 1998).

Wildfires occur throughout the year. The 2000 wildfire season began with a Florida fire on January 1, continued with two 40,000-acre fires in New Mexico, an early May, 47,000-acre fire near Los Alamos and peaked on August 29, 2000, when fires that eventually burned 1,642,579 acres were burning in 16 States (NIFC 2001a). Generally, the occurrence of wildfires moves northward from the Southeastern and Southwestern States as summer approaches, fuels dry and fire danger increases.

Wildfires, both in number and total acreage burned, vary widely from year to year and from region to region. Figures 3-5 and 3-6 show no consistent relation between the number of fires and acres burned. It is known, however, that smoke from these fires impacts air quality on both an episodic and long-term average basis over wide regions.

Wildfires occur as episodic events. For example, in 1999, smoke from fires reduced visibility to less than 100 feet (30 m) in Florida, prompting officials to advise people with respiratory problems to stay indoors (New York Daily News 1999). In the West, fires in six States (California, Nevada, Oregon, Montana, Washington, and Idaho) put thick smoke in many communities. In Reno and cities in California's Central Valley, smoke from nearby wildfires prompted authorities to warn residents with asthma to avoid unnecessary activity (USA Today 1999). Wildfire smoke is also transported across international boundaries. Fires in Canada were found to cause high concentrations of carbon monoxide and ozone over a period of 2 weeks in the Southeastern United States and across the Eastern seaboard during the summer of 1995 (Wotawa and Trainer 2000).

Smoke impacts during these episodic events can threaten public health, cause smoke damage to buildings and materials, and disrupt community activities. Although particulate concentrations in ambient air rarely reach health-threatening levels within major cities, several communities in the United States have experienced particulate matter concentrations from wildfire smoke that exceeded the Environmental Protection Agency (EPA) significant harm emergency action level of 600 $\mu g/m^3$ defined as an "imminent and substantial endangerment of public health" (EPA 1992b).

For example, the Yellowstone National Park wildfires of 1988 impacted communities in three States. Concentrations of suspended particulate matter — both total suspended particulate (TSP) and PM10 — measured in communities near the fires exceeded NAAQS, triggering public health alerts and advisories (Core 1996). An estimated 200,000 people were exposed to high concentrations of smoke. In 1987, the Klamath fires of northern California burned for more than 60 days, resulting in widespread smoke intrusions into numerous communities in northern California and southern Oregon. More recently, wildfire impacts during the 2000 season were also severe in several communities. Twenty-four average PM10 concentration measured in Salmon, ID, reached 225 $\mu g/m^3$ on August 15, 2002, and 281 $\mu g/m^3$ on August 18, 2000, during wildfire smoke intrusions (Idaho Department of Environmental Quality n.d.).

Wildfire smoke can also be the dominant cause of visibility reduction during episodic events in the Rocky Mountain States, on the Pacific Coast, and in the Southeast (National Research Council [NRC]

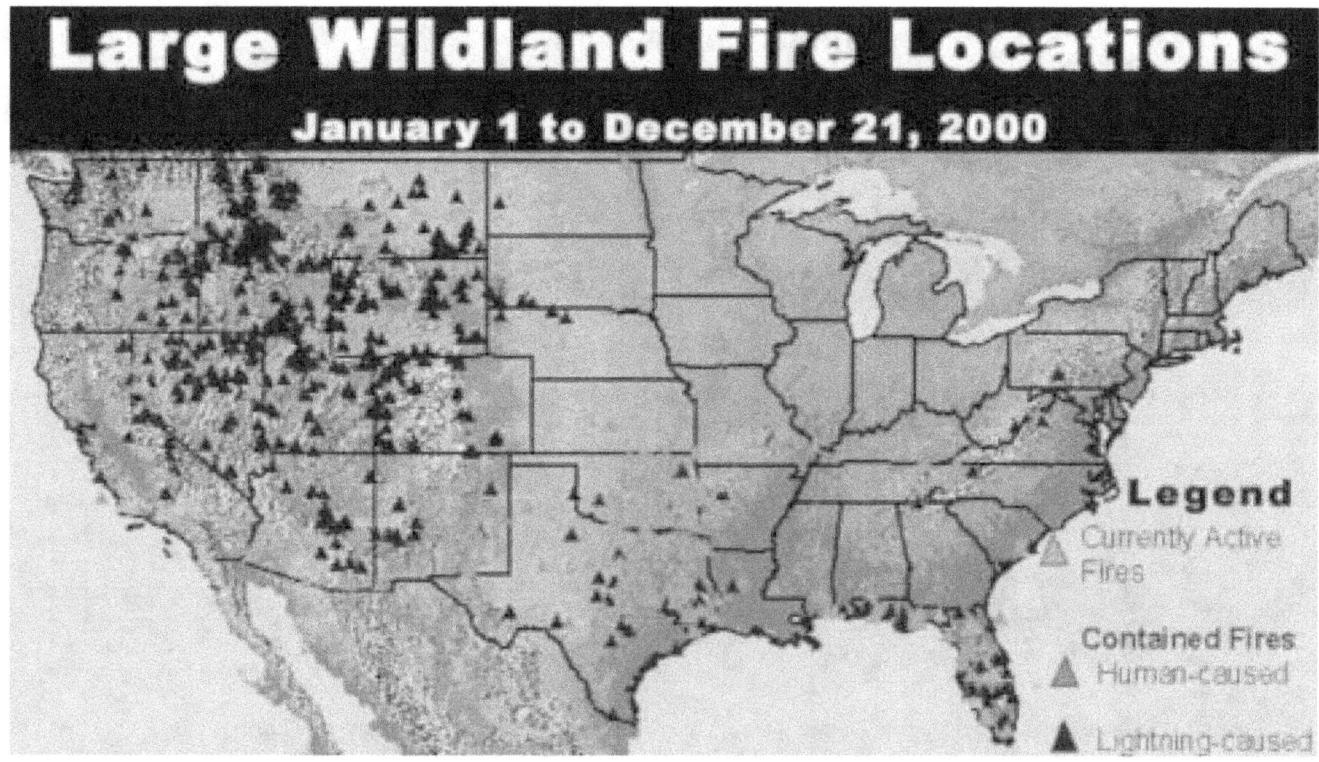

Figure 3-3—Location of major wildfires in 2000 available at http://www.nifc.gov/fireinfo/2000/Top10fires.html.

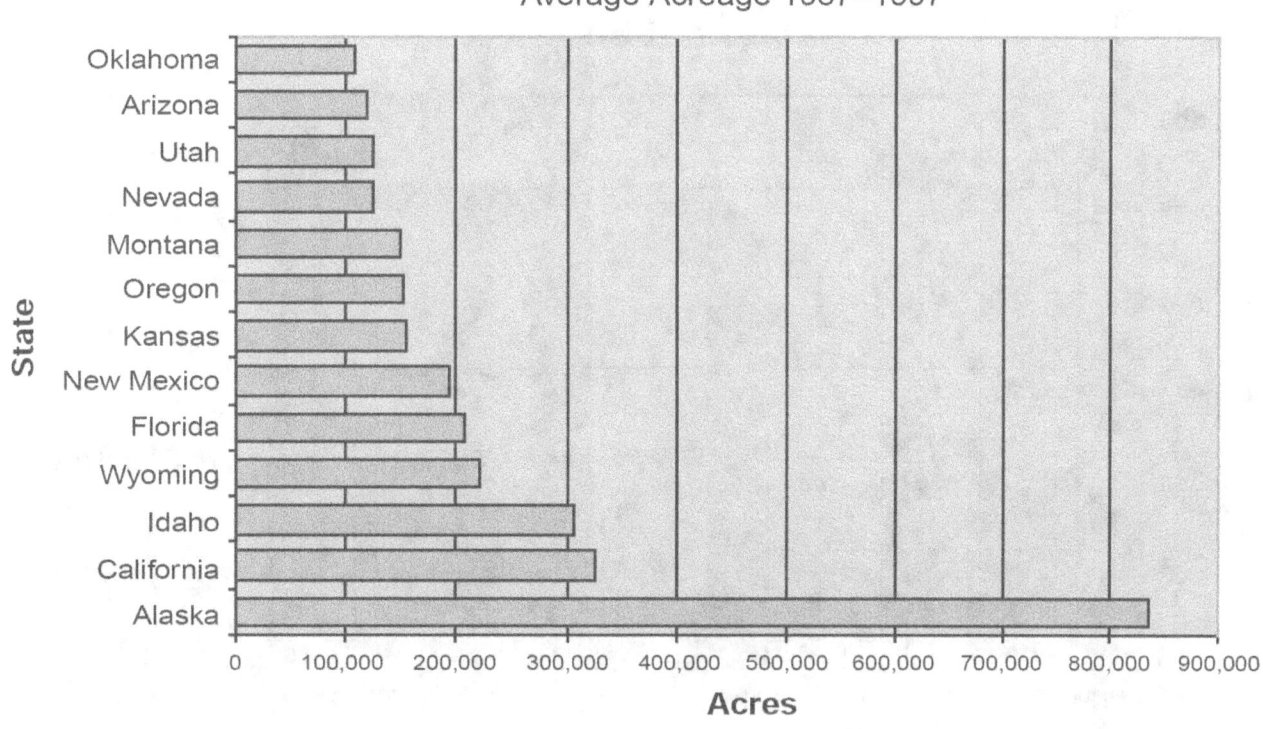

Figure 3-4—States with more than 100,000 acres per year burned by wildfires.

USDA Forest Service Gen. Tech. Rep. RMRS-GTR-42-vol. 5. 2002

21

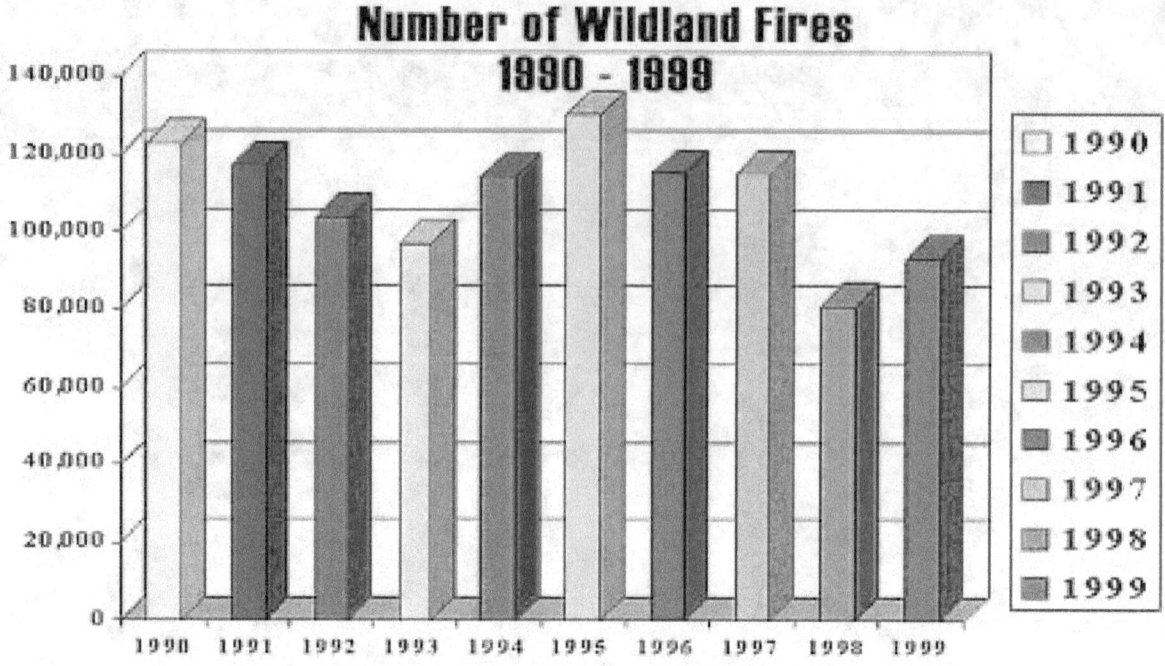

Figure 3-5—Number of wildfires per year 1990 through 1999 (National Interagency Fire Center 2002).

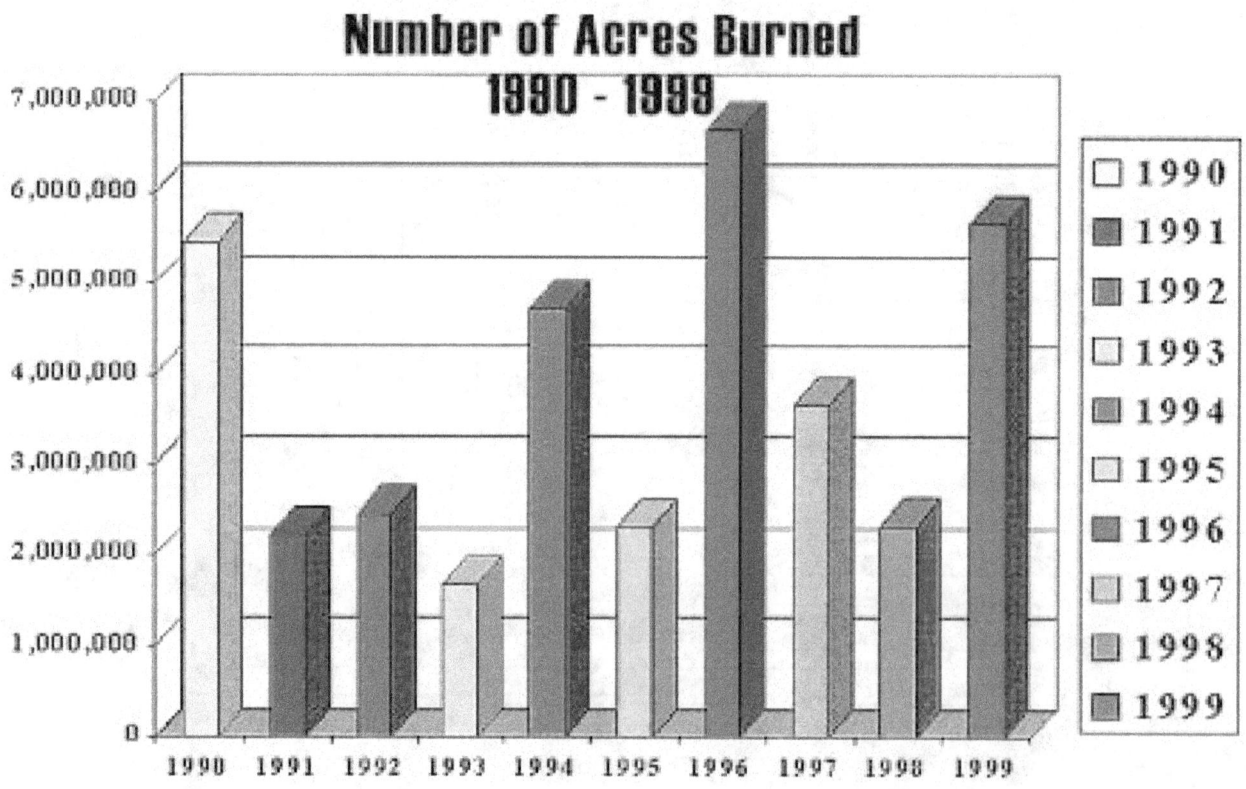

Figure 3-6—Number of acres burned by wildfires per year 1990 through 1999.

1993). Figures 3-7 and 3-8 are examples of the dense plumes of smoke that can be transported over hundreds of kilometers across State and international boundaries, degrading air quality, scenic values, and highway safety. Between 1979 and 1988, 28 fatalities and more than 60 serious injuries were attributed to smoke that drifted across roadways in the Southern United States (Mobley 1989).

Figure 3-7—Big Bar Fire, Shasta-Trinity National Forest, California, August 1999 (National Interagency Fire Center 2000).

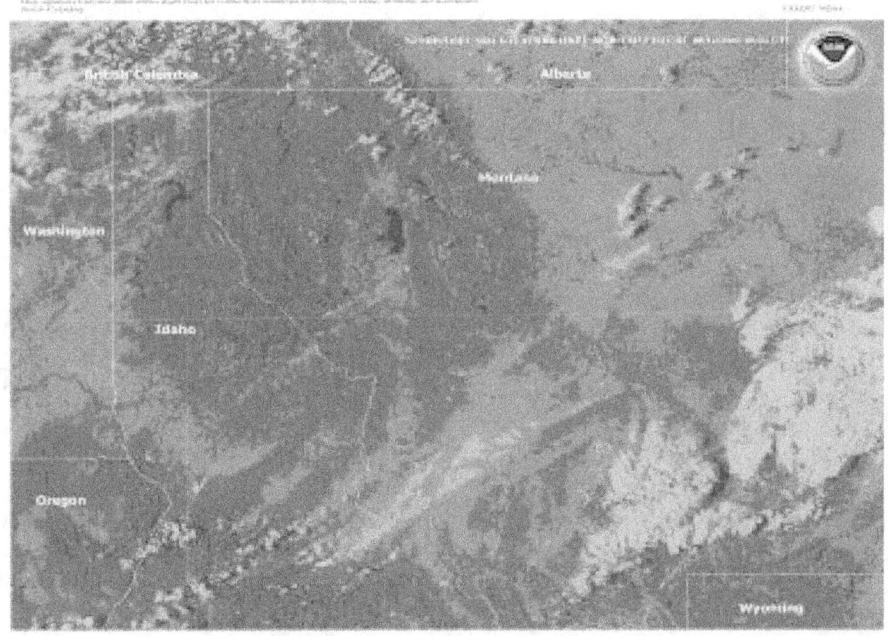

Figure 3-8—Wildfire smoke transported across State lines, August 14, 2000 (NASA).

USDA Forest Service Gen. Tech. Rep. RMRS-GTR-42-vol. 5. 2002

23

Smoke from Prescribed Fires

On a national annual basis, PM10 emissions from prescribed burns in 1989 were estimated to be over 600,000 tons, half of which (380,000 tons) occurred in the Southeastern States. Of the remaining 42 States, seven (Arizona, California, Idaho, Montana, Oregon, Texas, and Washington) were estimated to have annual emissions over 10,000 tons of PM10 from prescribed forest and rangeland burning (EPA 1992a; Peterson and Ward 1990). More recent estimates of prescribed fire PM2.5 emissions in the West (EPA regions 8, 9 and 10) totaled 193,293 tons (Dickson and others 1994). These national, annual estimates are less significant in terms of air quality impact than those prepared at the State level. For example, the 211,000 tons of prescribed fire PM10 emissions in Georgia in 1989 is about 30 percent of the total estimated particulate inventory for all sources (EPA 1992a). On a seasonal basis, emissions from prescribed burning are likely to be an even more significant percentage of total emissions in some States.

Acreage treated by prescribed burning on Federal lands increased from 918,300 acres in 1995 to 2,240,105 acres in 1999, demonstrating renewed interest in the use of fire as an important tool in the management of wildlands (NIFC 2001b).

Impacts on National Ambient Air Quality Standards

Characterization of the true extent of effects of prescribed and wildland fires on ambient air quality is incomplete due to the deficiency of air quality monitoring sites in rural areas. Also, particulate standards are based on 24-hour and annual averages, whereas smoke plumes may significantly degrade air quality in a community for just a few hours before moving or dispersing. These short-term, acute impacts likely cause discomfort at the least, and possibly even affect health, but may not result in a violation of the NAAQS.

Numerous exceedances of 24-hour PM10 and PM2.5 standards have been attributed to wildfires but, as mentioned previously, violations of NAAQS caused by wildfire do not result in nonattainment if a State can document that the cause of the violation was truly wildfire and then prepares a natural events action plan for future events.

At present, prescribed fires are not considered to be a significant cause of nonattainment, but with increased burning to reduce fuels, this situation may change as land managers move forward with implementing a several-fold increase in the use of fire to sustain ecosystems (USDI and USDA 1995; USDA 1997). In general, little information is available on a national level to identify the contribution of prescribed burning to PM10 or PM2.5 within nonattainment areas (EPA 1992a). It appears, however, that there is no clear relation between total acres burned (or particulate emissions) and the nonattainment status of nearby airsheds, possibly because of successful smoke management programs.

In areas where air quality standards are being or may be violated, however, land managers are being directed to reduce air quality impacts through smoke management programs. This is because any source that contributes even a few micrograms per cubic meter of particulate matter toward violation of the NAAQS may be required to reduce emissions to assure that air quality standards are attained.

Significance of Visibility Degradation

As noted above, wildland fires can significantly degrade visibility during episodic events. With the new emphasis on the reduction of regional haze in the Class I National Parks and wilderness areas of the nation, smoke from fire is of special concern, especially in the West. In their report to the EPA, the Grand Canyon Visibility Transport Commission (GCVTC) noted that emissions from fire, both wildland fire and prescribed fire, are likely to have the single greatest impact on visibility at Class I areas through 2040. During periods of intense fire activity, smoke from wildland fires is likely to make the worst 20 percent of days at the Grand Canyon even worse rather than impair visibility on clear days (GCVTC 1996b). The Commission recommended several actions to reduce impacts on regional haze including enhanced smoke management programs and establishment of annual emissions goals for all fire programs.

Greenhouse Gas Emissions from Fires

Globally, fires are a significant contributor of carbon dioxide and other greenhouse gases in the atmosphere. Fires account for approximately one-fifth of the total global emissions of carbon dioxide (Levine and Cofer 2000; Schimel 1995). Andreae and Merlet (2001) calculate that 5,130 Tg per year of biomass is consumed in fires, emitting 8,200 Tg per year of carbon dioxide, 413 Tg per year of carbon monoxide, and 19.4 Tg per year of methane. The accuracy of these global estimates is thought to be within plus or minus 50 percent, with the bulk of the error resulting from inaccuracies in the estimates of the area burned and the mass of fuel consumed.

Fires in temperate ecosystems are minor contributors compared to the world's savannas, boreal forests, and tropical forests. More than 60 percent of the totals listed in the previous paragraph are released from savannas and grasslands, and another 25 percent from tropical forests. Burning in tropical Africa is dominated by savanna fires; in tropical Asia, by forest

24

USDA Forest Service Gen. Tech. Rep. RMRS-GTR-42-vol. 5. 2002

fires; and in tropical South America, about equally represented by savannas and tropical forests (Hao and Liu 1994). Lavoué and others (2000) detail contributions from temperate and boreal fires, demonstrating that about 90 percent of the global boreal fire area is in Russia and Canada. Alaska accounts for only about 4.5 percent of the global boreal forest, but it accounts for at least 10 percent of the emissions from that source, because of the heavier fuel loads in Alaska. Alaska accounts for an average of 41 percent of total U.S. fire emissions, with a huge year-to-year variability. In 1990, 89 percent of U.S. fire emissions were from Alaska fires.

Smoke Management Programs

Smoke management programs establish a basic framework of procedures and requirements when managers are considering resource benefits. These programs are typically developed by States and Tribes with cooperation and participation by wildland owners and managers. The purposes of smoke management programs are to mitigate the nuisance (such as impacts on air quality below the level of ambient standards) and public safety hazards (such as visibility on roads and airports) posed by smoke intrusions into populated areas; to prevent significant deterioration of air quality and NAAQS violations; and to address visibility impacts in Class I areas.

The *Interim Air Quality Policy on Wildland and Prescribed Fires* (EPA 1998) provides clear guidelines for establishing the need for and content of smoke management programs and assigns accountability to State and Tribal air quality managers for developing and adopting regulations for a program. Measured PM10 NAAQS exceedances attributable to fires, including some prescribed fires and wildland fires managed for resource benefits, can be excluded from air quality data sets used to determine attainment status for a State. Special consideration will be given if the State or Tribal air quality manager certifies in a letter to the administrator of EPA that at least a basic smoke management program has been adopted and implemented.

States with smoke management programs that have authorized a central agency or office to make burn/no-burn decisions include Arizona, Colorado, Oregon, Idaho/Montana, Washington, California, Nevada, New Mexico, Florida, South Carolina, Utah, North Carolina, and Wyoming (Battye and others 1999). In many other States, the decision to burn rests in the hands of the persons conducting the burn, local fire departments, or local authorities. These States include Alaska, Alabama, Arkansas, Georgia, Louisiana, Mississippi, Tennessee, Texas, and Virginia. In yet other States

(New York, Illinois, Massachusetts, and others), burn permits are required and may be subject to State air agency oversight if burning is conducted near nonattainment areas or areas sensitive to smoke (Core 1998; Hardy and others 2001). In addition, many private landowners, nonprofit conservation organizations and government agencies voluntarily practice responsible smoke management to maintain goodwill in their communities.

Smoke management programs have been established and are operated on an on-going basis because of local, regional, and national concerns about the impact of prescribed burning on air quality. The number, complexity, and cost of operating these programs underscore the potential significance of prescribed fire's impact on air quality on a national scale.

Smoke management programs across the nation have changed significantly since the mid-1980s. In the Pacific Northwest, there have been reductions in prescribed fire smoke management programs because of the decline in large-scale clearcut burning of forest harvesting residues. Current smoke management programs across the West have to place a much greater focus than in the past on understory burning to restore declining forest health, on burns to reduce fire hazards, or on burns to meet wildlife habitat objectives. All across the nation, an increasing number of people living within the wildland-urban interface have placed new emphasis on the need to minimize smoke impacts on residents living near fires. Increasing air quality regulatory pressures, fire manager liability issues, and the increased likelihood of fire escapement in overstocked forestlands have all placed ever-greater demands on fire practitioners.

As these demands have increased, so have the number and complexity of smoke management programs nationwide (Hardy and others 2001). Although the complexity of these programs varies widely from State to State, the key to a successful program always lies in its ability to balance the use of prescribed fire with air quality, environmental, legal, and social requirements. Increasingly, this has meant adoption of formalized burn authorization procedures issued by program managers who are responsible for overseeing burning on both public and private lands on a daily basis. Coordinated burn operations are based on meteorological forecasts, the location of smoke-sensitive receptors, fuel conditions, and a myriad of other considerations. Increasingly, public notification of planned burning activity and monitoring of smoke transport, as well as fire practitioner training and program enforcement, are becoming more common (Battye and others 1999).

As inter-State smoke transport becomes a larger issue, agencies are expanding coordination. For example, land management agencies in California's

USDA Forest Service Gen. Tech. Rep. RMRS-GTR-42-vol. 5. 2002

25

San Joaquin Valley are using a new centralized, electronic database, Prescribed Fire Incident Reporting System (PFIRS), to schedule fires and to share information on expected emissions and smoke transport with California and Nevada air and land management agencies (Little n.d.). This trend is likely to continue as States begin to work on regional haze control programs.

The Western Regional Air Partnership (WRAP) Fire Emissions Joint Forum (FEJF) has issued a draft policy to set the criteria for enhanced smoke management plans for visibility protection in the West (Fire Emissions Joint Forum 2002). The policy document concludes that the regional haze rule can be satisfied only by the States and Tribes establishing an emission tracking system for all prescribed fires and wildland fires; by managing smoke from all fires; and by implementing smoke management systems that include nine elements:

1. Actions to minimize emissions from fire
2. Evaluation of smoke dispersion
3. Alternatives to fire
4. Public notification of burning
5. Air quality monitoring
6. Surveillance and enforcement
7. Program evaluation
8. Burn authorization
9. Regional coordination

The enhanced smoke management plan (ESMP) policy would enable Western States and Tribes to minimize increases in emissions and show reasonable progress toward the natural visibility goal. The Fire Emissions Joint Forum is developing additional policy and technical tools that will support ESMP policy and its implementation, such as recommendations for creation of an annual emissions goal, availability and feasibility of alternatives to burning, recommendations for managing fire emissions sources, guidance for feasibility determinations, and a method for tracking fire emissions.

26

USDA Forest Service Gen. Tech. Rep. RMRS-GTR-42-vol. 5. 2002

Chapter 4: Characterization of Emissions from Fires

All fires emit air pollutants in addition to nonpolluting combustion products; but fires vary widely in what pollutants are emitted in what proportion. Characterizing and managing air pollution from fires first requires knowledge of the amount and timing of what pollutants are emitted. Fires are a complex combustion source that involve several stages of combustion, several categories of fuels, and fire behavior that changes over time and with fuel and weather conditions; so the amount, rate, and nature of pollutants also vary widely. Characterizing emissions from fires requires explicit knowledge of fuelbed character and condition, combustion environment, and fire behavior.

This chapter reviews the state of knowledge and predictive models necessary to characterize air pollutant emissions from prescribed and wildland fires.

All components of smoke from fires, with the exception of carbon dioxide and water, are generated from the inefficient combustion of biomass fuels. The amount of smoke produced is derived by determining the fuel consumed (tons per acre) in each combustion stage and knowing the size of the area burned, fuel characteristics, fire behavior, and combustion conditions (fuel moisture, weather parameters, and so forth). The fuel

consumption is then multiplied by an emissions factor for each pollutant, which is an expression of the efficiency of combustion. An emission factor is the ratio of the mass of pollutant per unit mass of fuel consumed, and is a statistical average of measurements made in the plumes of fires containing differing fuel types and combustion stages. Errors and uncertainties arise in the estimates made during each step in the process of estimating emissions.

Area Burned

At first glance, amount of area burned seems relatively easy to calculate. However, individual estimates of fire size tend to be systematically exaggerated, and fires are frequently double-counted in inventories. For example, geographic features, nonuniform fuelbeds, or a change in the weather will often cause a fire to create a mosaic of burned, partially burned, and unburned areas, although the entire landscape within the fire perimeter is often reported as burned. In addition, large-scale (such as continental) inventories of area burned are often derived from remote sensing data that have resolutions from 250 m to 1 km (SAI 2002), limiting their precision. Remote

USDA Forest Service Gen. Tech. Rep. RMRS-GTR-42-vol. 5. 2002

27

sensing accuracy is currently inadequate in landscapes that change slope and fuel characteristics over a few tens of meters.

Preburn Fuel Characteristics _____

Large variations in fuel characteristics can contribute up to 80 percent of the error associated with predicting emissions (Peterson 1987; Peterson and Sandberg 1988). Fuel characteristics can vary widely across the landscape (figs. 4-1 and 4-2). For instance, fuel loads can range from less than 3 tons per acre for perennial grasses with no rotten woody material or duff, 6 tons per acre in a sagebrush shrubland, 60 tons per acre in a ponderosa pine and Douglas-fir forest with rotten woody material, stumps, snags, and deep duff, to 160 tons per acre in a black spruce forest with deep moss and duff layer. The greatest errors occur when the fuel load is inferred from vegetation type as

is usual when deriving biomass emissions from remotely sensed data (Crutzen and Andrae 1990; Levine 1994). Preburn fuel characteristics, such as relative abundance for particular fuelbed components (grasses, shrubs, woody fuels, litter, duff, and live vegetation) and the condition of the fuel (live, dead, sound, rotten) are needed to calculate fuel consumption, and the resulting smoke.

The ongoing development of several techniques, including the natural fuels photo series (Ottmar and Vihnanek 2000a) and the fuel characteristic classification (FCC) system (Sandberg and others 2001), will provide managers new tools to better estimate fuel loadings and reduce the uncertainty that currently exists when assigning fuel characteristics across a landscape. The photo series is a sequence of single and stereo photographs with accompanying fuel characteristics. The FCC is a national system designed for classifying wildland fuelbeds according to a set of

Figure 4-1—Fuelbed types and fuel loads (a) grassland (3 tons per acre), (b) sagebrush (6 tons per acre), (c) ponderosa pine with mortality in mixed fir (60 tons per acre), and (d) black spruce with deep duff and moss (160 tons per acre). (Photos by Roger Ottmar)

Figure 4-2—Various fuelbeds across a single landscape. (Photo by Roger Ottmar)

inherent physical properties, thereby providing the best possible fuels estimates and probable fire parameters based on available site-specific and remotely sensed information.

Fire Behavior

Fire behavior is the manner in which fire reacts to the fuels available for burning (DeBano and others 1998) and is dependent upon the type, condition, and arrangement of fuels, local weather conditions, topography, and in the case of prescribed fire, ignition pattern and rate (fig. 4-3). Important aspects of fire behavior include:

- Fire intensity (rate of energy release per unit area or unit length of fire perimeter, generally during the flaming combustion period).
- Rate of spread (rate of advancement of flaming front, length per unit time), crowning potential (involvement of tree and shrub foliage and spread within the canopy), smoldering potential (smoldering combustion of fuels that have been preheated or dried during the flaming stage).
- Residual smoldering potential (propagation of a smoldering combustion front within porous fuels such as rotten logs or duff, independent of preheating or drying).
- Residence time in the flaming, smoldering, and residual stages of combustion.

These aspects influence combustion efficiency of consuming biomass, as well as the resulting pollutant chemistry and emission factor (fig. 4-4).

The Emissions Production Model (EPM) (Sandberg 2000; Sandberg and Peterson 1984) and FARSITE (Finney 1998) take into account fire behavior and ignition pattern to estimate emission production rates. Fire behavior during the flaming stage of combustion in surface woody fuels and some shrub vegetation is effectively predicted within models such as BEHAVE (Andrews and Bevins 1999) and its spatial application, FARSITE (Finney 1998). However, EPM and other applications do not consider fire intensity or other fire behavior attributes when estimating emissions from flames, and that may result in a reasonable approximation for criteria pollutants but also be a limitation to the estimate of hazardous air pollutants or trace gases. BURNUP (Albini and Reinhardt 1997), FARSITE (Finney 1998), and EPM v2.0 (Sandberg 2002) attempt to model the extent and duration of flaming and smoldering combustion in downed woody fuels and duff. Current capability to model residual combustion, combustion in rotten logs and duff, and fire behavior in the foliage canopies of trees and some shrubs remains inadequate to predict emission rates with any reasonable degree of accuracy.

The Los Alamos and Lawrence Livermore national laboratories offer an approach to predicting fire behavior, plume trajectory, and dispersion, by combining a fire physics model, FIRETEC, with a dynamic atmosphere model, HIGRAD, to produce a highly detailed numerical simulation of fire spread and atmospheric turbulence (Bradley and others 2000). The approach builds on prior experience in predicting the dispersion of hazardous air pollutants from fires such as burning oil fields or "nuclear winter" scenarios. This modeling approach is limited to the propagating front but is unique in its coupling of atmospheric and fire physics.

Figure 4-3—Fire behavior in the leaf layer of a longleaf pine forest. (Photo by Roger Ottmar)

USDA Forest Service Gen. Tech. Rep. RMRS-GTR-42-vol. 5. 2002

29

Figure 4-4—Fuel consumption in (a) large rotten log during a fall prescribed burn, (b) pile burning during a prescribed burn, (c) litter and duff during a prescribed burn, (d) grass during a wildfire, and (e) sagebrush during a prescribed fire. (Photos by Roger Ottmar)

Combustion Stages

At least three important stages of combustion exist when fuel particles are consumed (Mobley 1976; NWCG 1985): flaming, smoldering, and residual (also known as "glowing," "residual smoldering," or "residual combustion") (fig. 4-5). The efficiency of combustion is distinct for each stage, resulting in a different set of chemical compounds and thermal energy being released at different rates into the atmosphere. In the flaming phase, combustion efficiency is relatively high and usually tends to emit the least amount of pollutant emissions compared with the mass of fuel consumed.

The predominant products of flaming combustion are CO_2 and water vapor. During the smoldering phase, combustion efficiency is lower, resulting in more particulate emissions generated than during the flaming stage.

Smoldering combustion is more prevalent in certain fuel types such as duff, organic soils, and rotten logs, and often less prevalent in fuels with high surface to volume ratios such as grasses, shrubs, and small diameter woody fuels (Sandberg and Dost 1990).

30

USDA Forest Service Gen. Tech. Rep. RMRS-GTR-42-vol. 5. 2002

Figure 4-5—Flaming, smoldering, and residual combustion stages during a fire. (Photo by Roger Ottmar)

Fuel Consumption

Fuel consumption is the amount of biomass consumed during a fire and is another critical component required to estimate emissions production from fire. Biomass consumption varies widely among individual fires depending on the fuelbed type, arrangement, and condition, weather parameters, and the way the fire is applied in the case of prescribed fire. As with fuel characteristics, extreme variations can be associated with fuel consumption resulting in an error contribution of 30 percent or more when emissions are estimated (fig. 4-6) (Peterson 1987; Peterson and Sandberg 1988).

Biomass consumption of woody fuels, piled slash, and duff in forested areas has become better understood in recent years (Albini and Reinhardt 1997; Brown and others 1991; Ottmar and others 1993; Ottmar and others [N.d.]); Reinhardt and others 1997; Sandberg 1980; Sandberg and Dost 1990). Consumption of forested crowns and shrublands are the least understood components of biomass consumption, and research is currently under way (Ottmar and Sandberg 2000) to develop or modify existing consumption equations for these fuel components. Equations for predicting biomass consumption in the flaming and smoldering combustion stages are widely available in two major software packages, Consume 2.1 (Ottmar and others [N.d.]) and the First Order Fire Effects Model (FOFEM 5.0) (Reinhardt and Keane 2000).

The residual stage differs from the smoldering stage in that the smoldering stage is a secondary process that occurs in fuels preheated or dried by flaming combustion, while residual is an independent process of propagation in a fuelbed unaffected by the flaming stage. This phase is characterized by little smoke and is composed mostly of CO_2 and carbon monoxide. All combustion stages occur sequentially at a point, but simultaneously on a landscape.

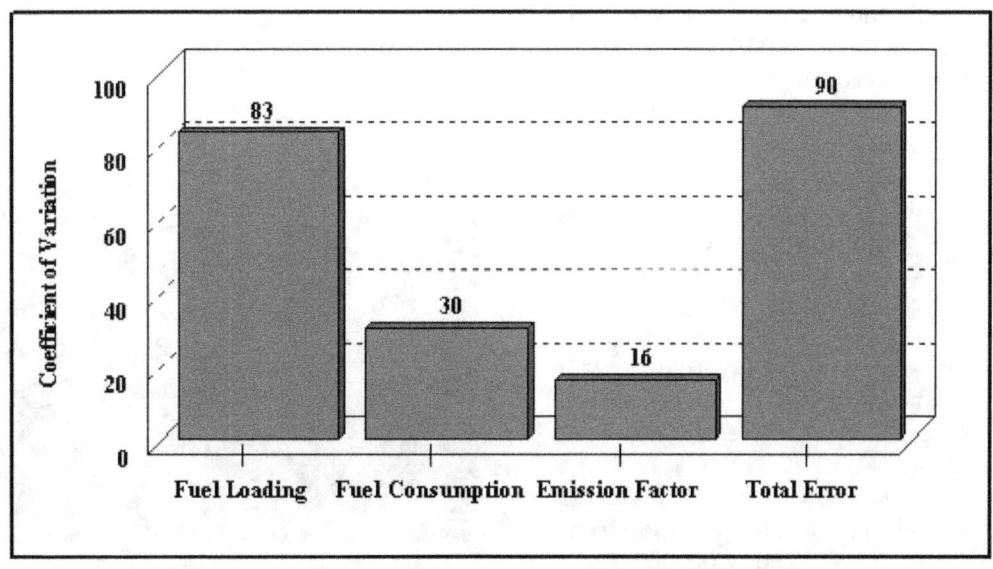

Figure 4-6—The largest errors are associated with fuel loading and fuel consumption estimates when determining emission production and impacts from wildland fire (Peterson and Sandberg 1988).

USDA Forest Service Gen. Tech. Rep. RMRS-GTR-42-vol. 5. 2002

31

Emission Factors

Emissions from fires or from points over fires have been observed extensively by researchers since about 1970. The result is a complete set of emission factors (pounds of pollutant per ton of fuel consumed) for criteria pollutants and many hazardous air pollutants for most important fuel types. These are available in several publications (for example, Battye and Battye 2002, EPA 1972, Hardy and others 2001, Ward and others 1989) and are not reproduced here.

Less complete compilations of emission factors are for particulate matter components such as size class distribution, elemental and organic carbon fractions, and particulate hazardous air pollutants; and for methane, ammonia, aldehydes, compounds of nitrogen, volatile organic hydrocarbons, and volatile hazardous air pollutants (for example, Battye and Battye 2002, Goode and others 1999, Goode and others 2000, Lobert and others 1991, McKenzie and others 1994, and Yokelson and others 1996).

Source Strength

Source strength is the rate of air pollutant emissions in mass per unit of time, or in mass per unit of time per unit of area. Source strength is the product of the rate of biomass consumption (that is, fuel consumption) and an emission factor for the pollutant(s) of interest and is representative of the physical and chemical fuel characteristics (fig. 4-7). Source strength or emission rate is required as an input to dispersion models (Breyfogle and Ferguson 1996), or to break down emission inventories into time periods shorter than the duration of a fire event. Source strength is also required in photochemical models such as the community multiscale air quality model (CMAQ) (Byun and Ching 1999) to account for timing of chemical reactions with diurnal patterns and interaction with other sources.

Total emissions from a fire or class of fires are the source strength integrated over the time of burning. Total emissions from a single class of fires (that is, a set of fires similar enough to be characterized by a single emission factor) can be estimated by multiplying that emission factor by the level of activity, which is the total biomass consumed by the class of fires. An emission inventory is the aggregate of total emissions from all fires or classes of fire in a given period for a specific geographic area.

Managing the source strength (or level of activity) of fires is the most direct way to control air pollution from wildland and prescribed fires. Prediction of source strength is sometimes used to manage the rate of emissions from fires, and it also is needed as an input to dispersion models. Standards or regulations are commonly set to limit the total emissions of pollutants, emission of specific hazardous air pollutants, or the level of activity, so that estimates of biomass consumption can be essential for environmental assessment, permitting of prescribed fires, or measuring compliance. Emission inventories are a critical part of impact analyses and strategy development so the level of activity must be estimated whenever there is a regulatory application.

The Emissions Production Model (Sandberg 2000; Sandberg and Peterson 1984) is currently the most widely used model for predicting source strength for prescribed fires. EPM v.1 predicts flaming and residual emissions rates for each criteria pollutant based on a simple formula that assumes a constant rate of ignition of a prescribed fire in uniform fuels. The software package pulls fuel consumption predictions from Consume 2.1 or FOFEM 5.0 and uses ignition pattern, ignition periods, and burn area components to calculate source strength for the flaming and residual combustion phases. EPM v.1 does not consider smoldering emissions (for example, long-duration, self-propagating glowing combustion), multiple fires or multiple burn periods, wildland fire or piled burning emissions, or diurnal and spatial changes in the fire environment. EPM v.2, now under development (Sandberg 2000), corrects all of these shortcomings in a dynamic simulation model. EPM v.2 will satisfy the requirement to provide hourly estimates of emission rates for most fires and fuelbeds needed for input into Models-3/CMAQ (see the "Grid Models" section in chapter 5) and into currently envisioned smoke management screening systems.

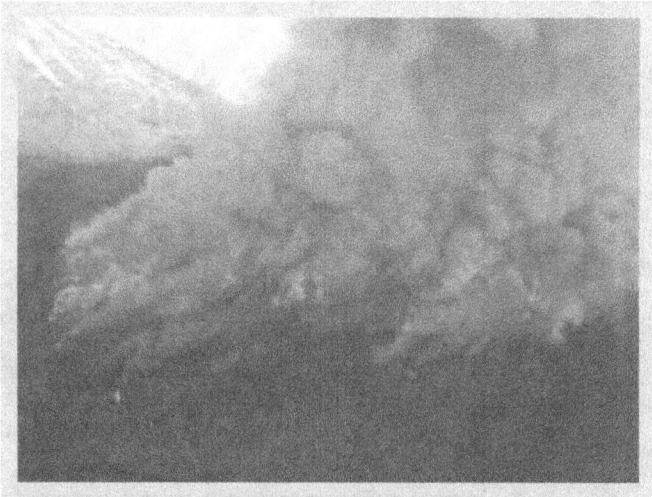

Figure 4-7—A high-intensity Alaska wildfire with heavy fuel loads, causes a high rate of emissions. (Photo by Roger Ottmar)

32

USDA Forest Service Gen. Tech. Rep. RMRS-GTR-42-vol. 5. 2002

FARSITE (Finney 2000) has been modified to predict emissions source strength as well as fire behavior in a detailed spatial simulation. FARSITE incorporates BURNUP (Albini and Reinhardt 1997), which estimates consumption and rates of individual fuel elements.

Accurate characterization of emissions from fires is critical to predicting the impact emissions will have upon communities and across broader landscapes and airsheds. Managers will increasingly be required to provide this type of information prior to prescribed burns, as well as during the course of wildland fires, and the information provided here summarizes the strengths and weaknesses of the various means of prediction.

USDA Forest Service Gen. Tech. Rep. RMRS-GTR-42-vol. 5. 2002

33

Notes

_____

_____

_____

_____

_____

_____

Chapter 5: Transport, Dispersion, and Modeling of Fire Emissions

To anticipate the impacts of smoke, the timing and location of smoke concentrations become important. Data on the site-specific surface concentrations of respirable particles and gases often are needed for estimating impacts on public health and welfare, requiring atmospheric dispersion and transport models that can approximate the atmospheric physics and chemical reactions that occur during transport near the ground. Data on the cumulative concentrations of elements that scatter and absorb light also are needed to estimate impacts on visibility and haze, requiring models that can approximate aqueous reactions as well as physical and chemical reactions at all levels of the atmosphere.

Although progress is being made, none of the currently available models fully meet the needs of fire planners and air resource managers. Much of the deficiency in current modeling approaches is caused by inherent uncertainties associated with turbulent motions between the fire, smoke, and the atmosphere that are compounded by the highly variable distribution of fuel elements, composition, and condition.

Another source of deficiency is that most available models were originally designed for well-behaved sources such as industrial stacks or automobile emissions, while emissions from fire can be extremely variable in both time and space. Also, outputs from currently available models do not always match the temporal or spatial scale needed for land management application.

To help readers understand the strengths and weaknesses of available models, we describe basic elements of the trajectory and dispersion of smoke. This chapter concludes with a summary of currently available models and a brief guide to applications.

Basic Elements of Trajectory and Dispersion

Ambient air quality can be measured at a point or as distribution of air quality over any space and time of interest. Ambient air quality is affected by the pollutants emitted to the atmosphere from fires, the background air quality that has already been degraded by other sources, the transport of the polluted parcels of the atmosphere, dispersion due to atmospheric movement and turbulence, secondary reactions, and removal processes. Plume rise is an important component of transport, because it determines where in the vertical structure of the atmosphere dispersion will begin.

USDA Forest Service Gen. Tech. Rep. RMRS-GTR-42-vol. 5. 2002

35

Overall, dispersion has proven extremely difficult to model accurately, especially in complex terrain. For example, detailed, gridded, three-dimensional meteorological data are required to model transport and dispersion, but expert judgment is often required to supplement or substitute for such modeled predictions.

Despite the difficulties of modeling, since about 1990 modeling systems used to assess the air quality impact of fires have grown increasingly important to both the fire planning and air quality communities. There is a broad range of acceptable tools from relatively simple methods used by local fire managers for estimating likely impacts on air quality standards (for example, SASEM: Riebau and others 1988; and VSMOKE: Lavdas 1996), to complex terrain and regional-scale models that incorporate atmospheric chemistry to assess impacts on regional haze (for example, Calpuff: Scire and others 2000a, and Models-3: Byun and Ching 1999).

The tremendous growth in model application places increasingly greater demands on the user, requiring access to detailed fuel characteristics, fuel consumption, ignition pattern, fire behavior, and meteorological inputs. Also needed is the ability to interpret the complex smoke dispersion model outputs.

In this section we describe such processes of heat release, plume rise and buoyancy, advection and diffusion, scavenging, and chemical transformations.

Heat Release

The consumption of biomass produces thermal energy, and this energy creates buoyancy to lift smoke particles and other pollutants above the fire. Heat release rate is the amount of thermal energy generated per unit of time. Total heat release from a fire or class of fires is a function of the heat content of the biomass, fuel consumed, ignition method and pattern, and area burned.

The early work of Anderson (1969) and Rothermel (1972) created fundamental equations for combustion energy in a variety of fuelbeds. Sandberg and Peterson (1984) adapted the combustion equations to model the temporal change in energy during flaming and smoldering combustion (Emission Production Model, EPMv.1.02). Currently, EPM provides heat release rates for most biomass smoke dispersion models (Harms and others 1997; Harrison 1995; Lavdas 1996; Sestak and Riebau 1988; Scire and others 2000a) and has been used to estimate the change in global biomass emissions patterns due to changes in land use (Ferguson and others 2000). The model, however, requires a constant rate of ignition with constant slope and wind. Such homogeneous conditions may be approximated during prescribed fires that are ignited with a deliberate pattern of drip torches or airborne incendiaries, or during portions of wildfires that experience relatively constant spread rates, both over fuelbed strata that retain a relatively consistent spatial and compositional pattern. To use EPM effectively for modeling source strength, the fire area and ignition duration are broken into space and time segments that meet the steady-state criteria.

Albini and others (1995), Albini and Reinhardt (1995), and Albini and Reinhardt (1997) do not explicitly derive temporal changes in combustion energy in their model, BurnUp, but they do assign source heat in steps of flaming and smoldering that are estimated from total fuel consumption. They have linked their model with the fire spread model, FARSITE (Finney 1998), which allows ignition rates and subsequent heat-release rates to vary over the landscape. The coupled system is computationally expensive and not yet associated with a plume rise component but may offer a reasonable approximation of the temporal and spatial varying emission rates of fires.

Plume Rise and Buoyancy

Heat, particle, and gas emissions from fires vary in time and space, causing unique patterns of convection and resulting plume rise. This plume rise is a function of free convection in the atmosphere, which is caused by density differences within the fluid. As a fire heats and expands air near the ground, large density differences between the heated volume and the surrounding air mass are created, causing the heated parcel to rise. The potential height of the resulting plume depends on the heat energy of the source and rise velocity, which is affected by the exchange and conservation of mass, radiant heat loss, the buoyancy force, and turbulent mixing with the ambient air.

Hot, flaming fires can develop central convective columns with counter-rotating vortices that involve massive entrainment of the surrounding air mass (Clark and others 1996; Haines and Smith 1987; Haines and Updike 1971). This stage of fire can produce fast-rising plumes and turbulent downdrafts, carrying sparks that ignite new fires. Cumulonimbus clouds often develop with accompanying lightning and rain. Dynamic plume rise brings gas and particles high into the atmosphere where strong winds can disperse the smoke hundreds to thousands of kilometers. As high intensity fires cool, however, the central column often collapses, creating numerous small convective cells that are less dynamic but equally active in carrying smoke into the atmosphere. Smoldering fires often create plumes that are neutrally buoyant, limiting widespread dispersion but allowing surface winds to dominate smoke trajectories. This can lead to accumulations of smoke in valleys and basins at night.

Because plume rise can eventually result in widespread dispersion, plume rise calculations are essential for determining the height above ground from which plume dispersion is initiated. Uncertainties in such calculations can result in inaccurate predictions of plume transport and downwind smoke impacts. Given the pressing need to predict the impact of plumes from fires, the need for improved plume rise calculations is apparent.

The basic mechanisms and algorithms used to describe plume rise and buoyancy were developed in the mid-1960s by Briggs (1969) for industrial, ducted emissions. These methods are still used today to estimate the plume rise and buoyancy of fires in spite of the significant differences in characteristics between ducted emissions and prescribed and wildland fires:

- Heat released from ducted sources is precisely known and usually emitted at relatively constant rates during a single phase of combustion. Heat released from fires is a function of fuel loading, fuel conditions, and ignition method through several phases of combustion (pre-ignition, flaming, smoldering, and residual), which create highly variable magnitudes and rates of heat release.
- Nearly all of the energy generated at the source of a ducted plume is transmitted to convection energy. In open burning, however, significant amounts of energy are lost by conduction and radiation, reducing the amount of available energy for convection.
- Plumes from ducted sources create single convective columns, but low intensity understory burning that occurs over broad areas does not develop a cohesive plume.

To improve plume rise predictions, emission production models need to do a better job of characterizing the spatial and temporal pattern of heat release from fires, and plume rise models need to be improved to account for the energy lost from the convective system through radiation and turbulent mixing. While models such as EPM and Burnup described in the previous section simulate variable rates of heat release from fires, both models use general estimates of spatial distributions of fuel, including structure, composition, and moisture content. Also, significant elements of fires that influence convective energy — such as the distribution of naturally piled fuel ("jackpots"), amount and density of rotten fuel and duff, and release of water vapor — are not adequately captured.

Rough approximations on the proportion of energy available for convection were made more than 40 years ago (Brown and Davis 1959). Despite efforts to improve plume rise calculations by removing the density difference assumption (Scire and others 2000a), they still are in use today.

Low intensity fires that typically do not have a cohesive convective column must be treated, from a modeling perspective, as an area source in Eulerian grid models. In Lagrangian dispersion models, there is currently no valid means of calculating plume rise from unconsolidated convection. Eulerian coordinates (used by box and grid models) are coordinate systems that are fixed in space and time, and there is no attempt to identify individual particles or parcels from one time to the next. Lagrangian models (bell-shape or Gaussian distribution pattern, often applied to plume and puff models) are used to show concentrations crosswind of the plume.

Another complication for modeling is that once plumes from fires enter the atmosphere, their fluctuating convection dynamics make them more susceptible to erratic behavior than well-mannered industrial stacks. For example, different parts of a plume can be carried to different heights in the atmosphere at the same time. This causes unusual splitting patterns if there is a notable wind shear between lofted elevations, causing different portions of the plume to be transported in different directions. Therefore, predictions of the plume's impact on visibility and air quality under these conditions become highly uncertain (Walcek 2002). Even when the behavior of plumes from fires resembles that of stack plumes, the varying and widely distributed locations of wildland sources prevent consistent study. For example, down-wash of plumes has been observed from ducted (stack) emissions after an inversion breaks up — conditions that are common at the end of an onshore breeze if the plume is above the inversion at its source (de Nevers 2000; Venkatram 1988) or if horizontal stratification in the lower atmosphere is disrupted by mountains (de Nevers 2000).

These characteristics of plumes from fire are strikingly different than those of ducted industrial emissions yet little research has been done on this topic in the past several decades.

Advection and Diffusion

In most existing models, the horizontal advection of smoke and its diffusion (lateral and vertical spread) are assumed to be controlled mainly by wind, and the formation and dissipation of atmospheric eddies. These elements are greatly simplified by assuming constant wind (at least for an hourly time step) in some cases (such as VSMOKE and SASEM), and a Gaussian dispersion is nearly always imposed. Perhaps the most critical issues are the constantly changing nature of the plume due to scavenging, chemical transformation, and changing convection dynamics that affect plume transport.

Many photochemical and dispersion models depend on gridded meteorological inputs. Unfortunately,

USDA Forest Service Gen. Tech. Rep. RMRS-GTR-42-vol. 5. 2002

37

numerical formulations of dynamic meteorological models (for example, MM5: Grell and others 1995; RAMS: Pielke and others 1992) do not adequately conserve several important scalar quantities (Byun 1999a, 1999b). Therefore, modelers often introduce mass-conserving interpolations. For example, Models-3/CMAQ (Byun and Ching 1999) uses the MCIP scheme (Byun and others 1999), Calpuff (Scire and others 2000a) employs CALMET (Scire and others 2000b), and TSARS+ (Hummel and Rafsnider 1995) is linked with NUATMOS (Ross and others 1988). Driving a photochemical or dispersion model without these mass-conserving schemes will produce inaccurate results, especially near the ground surface.

Scavenging

Smoke particles by nature of their small size provide efficient cloud condensation nuclei. This allows cloud droplets to condense around fine particles, called nucleation scavenging. Scavenging within a cloud also can occur as particles impinge on cloud droplets through Brownian diffusion, inertial impaction, or collision by electrical, thermal, or pressure-gradient forces (Jennings 1998). Cloud droplets eventually coalesce into sizes large enough to precipitate out, thus removing smoke aerosols from the atmosphere. While interstitial cloud scavenging, especially nucleation scavenging, is thought to dominate the pollution removal process, particles also may be removed by impacting raindrops below a cloud. Jennings (1998) reviews several theories on pollution scavenging but contends that there is little experimental evidence to support such theories.

The size and chemical structure of particles determine their efficiency in nucleation or other scavenging mechanisms. While the chemical composition of smoke is reasonably well known (see chapter 6), distributions of particle size from fire are not. The few airborne measurements (Hobbs and others 1996; Martins and others 1996; Radke and others 1990) do not distinguish fire characteristics or combustion dynamics, which play important roles in the range of particle sizes emitted from a fire. Therefore, the efficiency of scavenging biomass smoke particles out of the atmosphere by cloud droplets, rain, or other mechanism has not been quantified.

Chemical Transformations

Chemical transformations provide another mechanism for changing particle and gas concentrations within a plume. Chemical transformation in the plume can be important in regional-scale modeling programs where sulfate chemistry and ozone formation are of interest (see chapter 6). Oxidation within the smoke plume causes a loss of electrons during chemical transforma-

tion processes, which increases polarity of a molecule and improves its water solubility (Schroeder and Lane 1988). This improves scavenging mechanisms by cloud and rain droplets. Chemical transformation rates depend on complex interactions between catalysts and environmental conditions such as turbulent mixing rates.

Transport and Dispersion Models

Trajectories show the path of air parcels along a streamline in the atmosphere. Their simplicity allows trajectory methods to be used as a diagnostic tool for identifying the origin of air parcels from a potential receptor. This commonly is called a backward trajectory or back trajectory analysis. Because these models integrate over time the position of a parcel of air that is transported by wind, their accuracy is limited by the grid resolution of the model. Also, the flow path of a single parcel may have little relation to an actual plume dispersion pattern.

Current models to predict trajectory or air quality impacts from fires are inadequate in coverage and are incomplete in scope (Sandberg and others 1999). But because of new interest in modeling emissions on a regional scale, land managers need transport and dispersion models that include all fire and fuel types as well as multiple sources. Such models need to be linked to other systems that track fire activity and behavior as well as provide for variable scaling to fit the area of interest. At the operational level, models that support real-time decisionmaking during fire operations in both wildland fire situation analysis and go/no-go decision making are also needed (Breyfogle and Ferguson 1996). Transport and dispersion models fall into four major categories. These categories include plume, puff, particle, and grid.

Plume Models

One of the simplest ways of estimating smoke concentrations is to assume that plumes diffuse in a Gaussian pattern along the centerline of a steady wind trajectory. Plume models usually assume steady-state conditions during the life of the plume, which means relatively constant emission rates, wind speed, and wind direction. For this reason, they can be used only to estimate concentrations relatively near the source or for a short duration. Their steady-state approximation also restricts plume models to conditions that do not include the influence of topography or significant changes in land use, such as flow from a forest to grassland or across a land-water boundary.

Gaussian plume models have a great benefit in places and circumstances that restrict the amount of available input data. They can be run fast and have simple but realistic output that can be easily interpreted. Many regulatory guidelines from the EPA are based on Gaussian plume models.

Plume models typically are in Lagrangian coordinates that follow particles or parcels as they move, assigning the positions in space of a particle or parcel at some arbitrarily selected moment. (Lagrangian coordinates are used by plume, puff, and particle models.) Examples adapted for wildland biomass smoke include VSMOKE (Harms and others 1997; Lavdas 1996) and SASEM (Riebau and others 1988; Sestak and Riebau 1988). Both models follow regulatory guidelines in their development and offer a simple screening tool for examining potential concentrations at receptor locations from straight-line trajectories relatively near the source. However, SASEM directly compares downwind concentrations with ambient standards and calculates visibility impairment in a simple manner. It is also used as a State regulatory model in Wyoming, Colorado, New Mexico, and Arizona, and has been recommended for use by the EPA.

Plume rise models developed for other applications might be useful if adapted to fire environments. For example, ALOFT-FT (A Large Outdoor Fire Plume Trajectory Model - Flat Terrain), developed for oil-spill fires (Walton and others 1996), is a computer-based model to predict the downwind distribution of smoke particulate and combustion products from large outdoor fires. It solves the fundamental fluid dynamic equations for the smoke plume and its surroundings with flat terrain. The program contains a graphical user interface for input and output, and a database of fuel and smoke emission parameters that can be modified by the user. The output can be displayed as downwind, crosswind, and vertical smoke concentration contours.

Puff Models

Instead of describing smoke concentrations as a steadily growing plume, puff models characterize the source as individual puffs being released over time. Each puff expands in space in response to the turbulent atmosphere, which usually is approximated as a Gaussian dispersion pattern. Puffs move through the atmosphere according to the trajectory of their center position. Because puffs grow and move independently of each other, tortuous plume patterns in response to changing winds, varying topography, or alternating source strengths can be simulated with some accuracy.

Some models allow puffs to expand, split, compact, and coalesce (Hysplit: Draxler and Hess 1998; Calpuff: Scire and others 2000a) while others retain coherent puffs with constantly expanding volumes (NFSpuff: Harrison 1995). In either case, the variability of puff generation, movement, and dispersion does not restrict the time or distance with which a plume can be

modeled. Most puff models are computed in Lagrangian coordinates that allow accurate location of specific concentrations at any time.

Particle Models

In a particle model, the source is simulated by the release of many particles over the duration of the burn. The trajectory of each particle is determined as well as a random component that mimics the effect of atmospheric turbulence. This allows a cluster of particles to expand in space according to the patterns of atmospheric turbulence rather than following a parameterized spatial distribution pattern, such as common Gaussian approximations. Therefore, particle models tend to be the most accurate way of simulating concentrations at any point in time. Because of their numerical complexity, however, particle models usually are restricted to modeling individual point sources with simple chemistry or sources that have critical components such as toxins that must be tracked precisely. Particle models use Lagrangian coordinates for accurate depiction of place of each time of particle movement (for example, Hysplit: Draxler and Hess 1998; PB-Piedmont: Achtemeier 1994, 2000).

Grid Models

Grid models use Eulerian coordinates, disperse pollutants uniformly within a cell, and transport them to adjacent cells. The simplicity of advection and diffusion in a grid model allows these models to more accurately simulate other characteristics of the pollution, such as complex chemical or thermal interactions, and to be used over large domains with multiple sources. This is why grid models commonly are used for estimating regional haze and ozone and are often called Eulerian photochemical models. Much of the future work on fire impact assessment and planning at regional to national scales will be done by using grid models.

Because of their nature, grid models are not used to define accurate timing or locations of pollutant concentrations from individual plumes, only concentrations that fill each cell. This means that sources small relative to the grid size, which create individual plumes, will introduce unrealistic concentrations in places that are outside of the actual plume. Ways of approximating plume position and its related chemical stage include nesting grids to finer and finer spatial resolutions around sources of interest (Chang and others 1993; Odman and Russell 1991), establishing nonuniform grids (Mathur and others 1992), and creating "plume-in-grid" approximations (Byun and Ching 1999; Kumar and Russell 1996; Morris and others 1992; Myer and others 1996; Seigneur and others 1983).

USDA Forest Service Gen. Tech. Rep. RMRS-GTR-42-vol. 5. 2002

39

Many regional haze assessments use the Regulatory Modeling System for Aerosols and Acid Deposition (REMSAD) (Systems Applications International 2002). This model was adapted from the urban airshed model–variable grid (UAMV) by removing its plume-in-grid feature and parameterizing explicit chemistry to improve computational efficiency. REMSAD incorporates both atmospheric chemistry and deposition processes to simulate sulfate, nitrate, and organic carbon particle formation and scavenging. As such, it is quite useful for simulations over large regions.

The Models-3/ CMAQ modeling system is designed to integrate the best available modules for simulating the evolution and dispersion of multiple pollutants at a variety of scales (Byun and Ching 1999). It includes chemical transformations of ozone and ozone precursors, transport and concentrations of fine particles and toxics, acid deposition, and visibility degradation.

At the other end of the grid modeling spectra are simple box models that describe pollution characteristics of a small area of interest. Box models instantaneously mix pollutants within a confined area, such as a valley. This type of model usually is restricted to weather conditions that include low wind speeds and a strong temperature inversion that confines the mixing height to within valley walls (Lavdas 1982; Sestak and others 1988). The valley walls, valley bottom, and top of the inversion layer define the box edges. The end segments of each box typically coincide with terrain features of the valley, such as a turn or sudden elevation change. Flow is assumed to be down-valley, and smoke is assumed to instantaneously fill each box segment. Few box models include the complex chemical or particle interactions that are inherent in larger grid models.

Model Application

Modeling of the transport and dispersion of industrial stack plumes has occurred for decades, prompting a variety of techniques. But application to fires is much more limited (Breyfogle and Ferguson 1996). Part of the reason for this is that source strength from undulating and meandering fires is so difficult to simulate accurately. Therefore, applications have been appropriate mainly for relatively homogeneous fuelbeds and steady state burn conditions. This has restricted most transport and dispersion modeling to fires on a local scale and to those started in harvest residue from land clearing operations where fuels are scattered uniformly over the landscape or collected into piles (Hardy and others 1993; Hummel and Rafsnider 1995; Lavdas 1996; Sestak and Riebau 1988). Global-scale modeling also has taken place where fuelbed and ignition patterns are assumed to be

approximately steady state in relation to the grid size (Kasischke and Stocks 2000; Levin 1996).

Gaussian plume models (Harms and others 1997; Lavdas 1996; Sestak and Riebau 1988; Southern Forest Fire Laboratory Personnel 1976) are useful for places with relatively flat terrain, for circumstances when input data are scarce, and for evaluating surface concentrations relatively near the source. These models typically require only an estimate of atmospheric stability, trajectory wind speed and direction, and emission rates. Fires are modeled independently. Therefore, accumulations of smoke from multiple fires are ignored. Some Western States require SASEM modeling of prescribed burns before they can be permitted (Battye and Battye 2002).

Puff models (Draxler and Hess 1998; Harrison 1995; Hummel and Rafsnider 1995; Scire and others 2000a) are needed when simulating long-range transport, or transport that occurs during changeable environmental conditions such as influences from complex terrain or variable weather. NFSpuff has an easy user interface, but because of its internal terrain data files it is restricted to applications in the Western States, excluding Alaska (Harrison 1995). Hysplit (Draxler and Hess 1998) currently is programmed to accept only 16 individual sources and assumes a constant rate of emissions with no plume rise. Hysplit (Draxler and Hess 1998) and Calpuff (Scire and others 2000a) both include simple chemistry. NFSpuff is the most commonly used puff model for prescribed fire planning (Dull and others 1998). All three models are linked to the MM5 meteorological model (Grell and others 1995). NFSpuff can function with a simple trajectory wind, and Hysplit and Calpuff can accept other gridded weather input data.

Particle models are used in coupled fire-atmosphere modeling (Reisner and others 2000) and for tracking critical signature elements (Achtemeier 1994, 2000; Draxler and Hess 1998). The sophistication of these types of models and their computational requirements, however, has thus far limited their application to research development or individual case studies.

Eulerian photochemical grid models are highly useful in estimating smoke concentrations from many sources over large domains. In addition, their ability to model secondary chemical reactions and transformations is needed for determining ozone concentrations and regional haze conditions. Regional planning organizations such as the Western Regional Air Partnership (WRAP), are evaluating the photochemical models Models-3/CMAQ (Byun and Ching 1999) and REMSAD (Systems Applications International 2002) for use in guiding State implementation plans (SIPs) and Tribal implementation plans (TIPs).

Additional work is needed to fill critical gaps in the modeling systems identified above. As the need for

40

USDA Forest Service Gen. Tech. Rep. RMRS-GTR-42-vol. 5. 2002

Chapter 6: Atmospheric and Plume Chemistry

Traditionally, ozone and secondary aerosol precursors have been discussed within the context of urban smog caused by auto exhaust and reactive organic compounds emitted from industrial facilities. But the same pollutant and tropospheric chemical reactions occur in both urban settings and in rural areas where wildfire smoke may be an important if not dominant source of ozone precursor emissions. In these situations, emissions from fire may play an important role in ozone formation as well as nitrate and, indirectly, sulfate aerosol formation, which results in visibility impairment and increased PM2.5 concentrations.

At present, there is an urgent need to understand the impact of fire emissions on emerging visibility and ambient air standards as they relate to fire planning at the strategic, programmatic, and operational scales (Fox and Riebau 2000; Sandberg and others 1999). Chemical processes that occur in plumes from fires, directly or indirectly, touch on a number of these issues and are critical to the development of a regional model that will be used to assess the impact of fire on air quality.

Because of the Environmental Protection Agency's (EPA) pressing regulatory need to assess inter-State ozone transport and sources of precursor emissions, a new regional-scale mechanistic model called Models-3/CMAQ (Byun and Ching 1999) is being used by the Ozone Transport Commission (OTC) region of North-eastern and Mid-Western States, and the Western Regional Air Partnership (WRAP). Future applications will likely involve regional haze modeling in other areas of the country. Oxides of nitrogen (NO_X) and volatile organic compounds (VOCs) emissions from fire in the OTC have not previously been considered significant, but the new model photochemistry module requires that precursor emissions be included for all sources. As Models-3/CMAQ develops, NO_X and VOC emissions from fire will be included in ozone and secondary modeling.

Ozone Formation in Plumes

Field observations of ozone formation in smoke plumes from fires date back nearly 25 years when aircraft measurements detected elevated ozone at the edge of forest fire smoke plumes far downwind (Stith and others 1981). More recent observations (Wotawa and Trainer 2000) suggest that high concentrations of ozone are found in forest fire plumes that are transported great distances and across international boundaries. Measurements made during EPA's 1995 Southern Oxidant Study indicate that Canadian forest fires changed the photochemical properties of air masses over Tennessee on days with strong fire influence. Regional background ozone levels were elevated by 10 to 20 ppb on fire impact days as compared with

USDA Forest Service Gen. Tech. Rep. RMRS-GTR-42-vol. 5. 2002

41

nonimpact days during the study. Aircraft measurements found that, although forest fire plumes were always well defined with respect to carbon monoxide, they gradually lost their definition with respect to ozone after being mixed into the boundary layer. The amount of ozone transported to the surface measurement sites was found to depend upon where and when the plumes reached the ground. Elevated plumes were always marked by enhanced ozone concentrations, at times reaching values of 80 to 100 parts per billion (ppb) above tropospheric background.

Stith and others (1981) mapped ozone mixing ratios in an isolated, fresh, biomass-burning plume. At the source, or near the bottom, of the horizontally drifting plume they measured low or negative changes in ozone values, which they attributed to titration by NO and low ultraviolet (UV) intensity. Near the top of the plume, 10 km downwind, and in smoke less than 1 hour old, they measured change in ozone values as high as 44 parts per billion by volume (ppbv). Greater changes in ozone were positively correlated with high UV. Thus the initial destruction of ozone by reactive species in the plume followed by its gradual formation was documented.

A new and potentially useful tool for assessing impacts of long-range plume transport is based on the concept of using $\Delta O_3/\Delta CO$ (excess O_3 over excess CO) as a "photochemical clock" to denote the degree of photochemical processing in a polluted air mass by using carbon monoxide as a stable plume signature. As the plume disperses, its volume expands and absolute values of ozone can drop even though net production of ozone is still occurring. The $\Delta O_3/\Delta CO$ normalizes for plume expansion and is a useful measure of net ozone production. In the course of atmospheric chemistry research, numerous observations of $\Delta O_3/\Delta CO$ ratios have been made in biomass burning haze layers. Unfortunately, the observations represent haze of various ages and uncertain origin. In haze layers 1 to 2 days old, changes in the $\Delta O_3/\Delta CO$ ratios of 0.04 to 0.18 were measured over Alaska (Wofsy and others 1992) and ratios of 0.1 to 0.2 were measured over Eastern Canada (Mauzerall and others 1996). High ratios, up to 0.88, were measured at the top of haze layers that had aged about 10 days in the tropics (Andreae and others 1994).

In 1997, airborne Fourier transform infrared spectroscopy (FTIR) measurements in large isolated biomass burning plumes in Alaska revealed new details of downwind chemistry. Downwind smoke samples that had aged in the upper part of one plume for 2.2 ± 1 hours had $\Delta O_3/\Delta CO$ ratios of 7.9 ± 2.4 percent, resulting from initial, absolute ozone formation rates of about 50 ppb/hr. Downwind samples obtained well inside another plume, and of similar age, did not have detectable ΔO_3, but did have $\Delta NH_3/\Delta CO$ ratios about one-third of the initial value. $\Delta HCOOH/\Delta CO$ (formic acid) and $\Delta CH_3COOH/\Delta CO$ (acetic acid) usually increased about a factor of 2 over the same time scale in samples from both plumes. NO_X was below the detection limit in all the downwind samples. These data provided the first precise in-plume measurements of the rate of O_3/CO increase and suggested that this rate depended on relative position in the plume. The apparently rapid disappearance of NO_X is consistent with the similar early observation, and the drop in NH_3 was consistent with a reaction with HNO_3 to form ammonium nitrate, which is a NO_X sink. Secondary sources of formic acid relevant to polluted air have been described (Finlayson-Pitts and Pitts 1986). Jacob and others (1992, 1996) discussed several gas-phase sources of acetic acid that could occur in biomass burning plumes. These experiments provide the first experimental indication of the approximate time scale of secondary organic acid production in actual plumes.

A large number of photochemical modeling studies of biomass burning plumes have been published (Chatfield and Delaney 1990; Chatfield and others 1996; Crutzen and Carmichael 1993; Fishman and others 1991; Jacob and others 1992, 1996; Koppmann and others 1997; Lee and others 1998; Lelieveld and others 1997; Mauzerall and others 1998; Olson and others 1997; Richardson and others 1991; Thompson and others 1996). Nearly all these studies conclude that the net production of ozone occurs either in the original plume, or as a result of the plume mixing with the regional atmosphere. Several studies have shown a strong dependence of the final modeled results on the details of the post-emission-processing scenario such as the timing between production of the emissions and their convection to the free troposphere (Chatfield and Delaney 1990; Jacob and others 1996; Lelieveld and others 1997; Pickering and others 1992; Thompson and others 1996).

Factors Affecting Plume Chemistry

The specific chemical composition of the plume depends on many factors: the details of post-emission atmospheric reactions including dilution rates, photolysis rates, position within the plume, altitude, and smoke temperature, which varies by time of day and combustion stage. Equally important is the chemistry of the downwind air that mixes with the plume, which could be clean air or contain aged plumes from urban areas or other fires. In addition, the physical aspects of the plume mixing are important. For example, at the relatively low temperatures typical of higher altitudes in the troposphere, peroxyacetyl nitrate (PAN) is a stable molecule, which can be transported. At lower altitudes, PAN can thermally decompose and rerelease NO_X. Nitric acid (HNO_3) can also be an important,

42

USDA Forest Service Gen. Tech. Rep. RMRS-GTR-42-vol. 5. 2002

transportable reservoir species for NO_X at high altitudes but for a different reason. HNO_3 has a narrower absorption cross-section at lower temperatures and therefore is less susceptible to photolysis. The rate of bimolecular reactions among smoke components usually decreases with temperature (thus typically with altitude or at night). Reaction rates depend even more strongly on the dilution rate, at least initially. Dilution by a factor of 2 will decrease a bimolecular reaction rate by a factor of 4.

Emission Factors for Reactive Species

Emission factors for hydrogen oxide (HO_X, a collective term for OH and HO_2) precursors, NH_3, and NO_X have been estimated with the Missoula, MT, open-path spectroscopic system (Yokelson and others 1997). These experiments reveal that smoke contains high levels of oxygenated organic compounds, methanol (CH_3OH), acetic acid (CH_3COOH), and formaldehyde (HCHO). These compounds typically oxidize or photolyze within hours in a smoke plume to release HO_X that is important in sulfate aerosol formation processes. Under clear-sky conditions typical for noon on July 1 at 40°N latitude, the formaldehyde photolysis lifetime is about 3.8 hours (Yokelson and others 1997). Since the HCHO/CO source ratio for fires is typically near 2 percent, this process clearly injects large quantities of HO_2 into fresh plumes (Yokelson and others 1997). HO_X emissions from fire may become a critical input to regional haze models that simulate secondary sulfate formation processes.

The H_2O_2 is soluble in cloud droplets where it would play a major role influencing reaction rates during aqueous-phase sulfate formation chemistry (NRC 1993).

Particle Formation in Plumes

A number of processes are important in plume particle formation and growth. Many of these processes involve interaction with the trace gases in a plume originating from nucleation in which two gases react to form a solid nucleus for subsequent particle growth. An example of nucleation is the reaction of ammonia and nitric acid. In addition, condensation can create new particles when gases cool or through particle growth when a trace gas collides with and condenses on an existing particle. The second condensation process is quite common because biomass burning aerosol is hydrated. Soluble nucleilike ammonium nitrate

promotes this process. There is a little evidence that organic gases also condense on particles. Nucleation and condensation are both examples of trace-gas-to-particle conversion, which will increase the mass of particles in a plume, decrease the concentration of certain trace gases in the plume, and, in the case of condensation, contribute to an increase in average particle diameter. Andreae and others (1988) measured particle-NH_4^+/CO_2 ratios of 0.7 to 1.5 percent in slightly aged biomass burning plumes. Measurements of NH_3/CO in fresh smoke are typically near 2 percent. Thus, there is probably rapid conversion of gas-phase NH_3 to particle NH_4^+ either through nucleation or dissolution in the surface water of other hydrated particles.

Coagulation is when two particles collide and combine. This increases the average particle diameter, reduces particle number, and does not effect total particle mass. Coagulation probably contributes to the increase in average particle diameter that occurs downwind from fires (Reid and others 1998).

At any given point in its evolution a particle may impact the trace gas chemistry in a smoke plume. For instance, it is known that NO_2 reacts on the surface of soot particles to yield gas phase HONO. This and other heterogeneous reactions such as ozone destruction may occur on smoke aerosol. Some recent research suggests that oxygenated organic compounds emitted from fires could also be important in heterogeneous processes. Hobbs and Radke (1969), Desalmand and others (1985), Andreae and others (1988), and Roger and others (1991) found that a high percentage (25 to 100 percent) of fire aerosol particles from fires could be active as condensation nuclei (CCN). Radke and others (1990) observed that cumulus clouds greater than 2 km in depth scavenged 40 to 80 percent of smoke particles. The high concentrations of CCN in smoke plumes can contribute to the formation of clouds with smaller than "normal" cloud droplet size distributions. This type of cloud is more reflective to incoming solar radiation and less likely to form precipitation. Some work suggests that absorbing aerosol can reduce cloud formation. Finally, clouds can evaporate and leave behind chemically altered particles.

All of these mechanisms alter both the chemical nature and number of particles contained within smoke plumes from fires. In addition, reactive species emitted from fires (see previous section) may alter the conversion rate of gaseous precursors of secondary sulfate and nitrate particles, affecting regional haze modeling results.

Although the regulatory implications of reactive species emissions from fire are yet to be determined, much more attention to these issues will occur once fire is including in regional haze and ozone modeling efforts.

USDA Forest Service Gen. Tech. Rep. RMRS-GTR-42-vol. 5. 2002

43

Notes

44

USDA Forest Service Gen. Tech. Rep. RMRS-GTR-42-vol. 5. 2002

Chapter 7: Estimating the Air Quality Impacts of Fire

State-of-the-science methods used to determine the impact of fire on air quality and visibility include: (1) emission inventories; (2) air quality monitoring instruments to measure smoke concentrations in real-time; and (3) filter-based monitoring techniques and receptor-oriented methods that quantify wildfire smoke contribution to air pollution based on the chemical characteristics of smoke particles or the spatial and/or temporal variability. Fire also contributes to ground-level ozone. These topics have become increasingly important to both air quality regulators and land managers as efforts to identify, or apportion, the contributions that fire makes to particulate air pollution, regional haze, and ground-level ozone come under increased scrutiny.

Because the health effects of air pollution are so difficult to measure in the broad population, there has been little effort to regulate or manage those effects directly. Many smoke management decisions are made on the basis of nuisance complaints as an indicator, rather than on quantitative measurements of impacts to health and welfare. Close to the source, efforts are being made to keep the exposure of firefighters to hazardous air pollutants within the standards set by the Occupational Safety and Health Administration. Hazard assessment describes the nature, concentration, and duration of pollutants. Exposure assessment quantifies the population exposed and the degree of exposure. Risk assessment describes the probable result for a population from all exposures. Integrated health risk assessments and economic assessments are still rare.

Modeling and data systems are needed to predict, measure, and monitor the ultimate effects of air pollution from fires on human or ecosystem health, on the economy, and on the comfortable enjoyment of life and property. Risk assessment methods are needed to compare these effects with those from other sources.

Emission Inventories

An emission inventory is an estimate of the mass of emissions by class of activity within a specified geographic area in a specified amount of time. Usually, an inventory is compiled by multiplying the appropriate emission factor (see chapter 4) by the estimated level of activity (in other words, tons of fuel consumed).

Development of emission inventory methods for fires was recently reviewed in detail by Battye and Battye (2002). The report considers prior attempts at emission inventory, describes approaches to estimating emissions from fires, and reviews the scientific information available as components of an inventory. The report also reviews emission reduction strategies and smoke management techniques.

USDA Forest Service Gen. Tech. Rep. RMRS-GTR-42-vol. 5. 2002

45

An emission inventory provides an understanding of the relative burden on the air resource from particular air pollution source categories. Emission inventories help explain the contribution of source categories to pollution events, provide background information for air resource management, provide the means to verify progress toward emission reduction goals, and provide a scientific basis for State air program development. An accurate emission inventory provides a measured, rather than perceived, estimate of pollutant production as the basis for regulation, management action, and program compliance. Emission inventories should include all important source categories including mobile, area, and stationary, and the inventories are not complete unless difficult-to-quantify sources such as agricultural burning, backyard burning, rangeland burning, and wildland and prescribed burning are addressed. Emission inventories are a basic requirement of State air resource management programs and are a required element of State implementation plans (SIPs). Emission inventories are also compiled annually at the national level and for specific geographic regions (sub-State, multi-State, or multi-jurisdiction) to address a particular regional air quality issue.

The science necessary to accurately estimate emissions from prescribed burning is quite good for most fuel types in the United States if good quality information about several critical variables is known. Area burned, fuel type, fuel loading, fuel arrangement, fuel consumption, and emission factors are all needed to accurately estimate emissions. Some of these require onsite reporting for reasonable accuracy including area burned, fuel type, and fuel arrangement. Other factors can be defaulted or estimated with reasonable accuracy if some other information is known. Fuel loading can be defaulted with knowledge of the fuel type and arrangement. Fuel consumption can be calculated with knowledge of the fuel type, fuel loading, and fuel moisture. Emission factor assignment is made with knowledge of the fuel type.

The science of predicting emissions from wildland fire is much weaker than for prescribed fire. In addition, it is generally far more difficult to obtain decent quality information about individual wildland fires.

In most cases, the information gap that makes fire emissions prediction a difficult endeavor is good quality, consistent, and regular reporting of the specific onsite variables needed for emissions estimation. Data collection systems that are supported and utilized by fire managers need to be developed for every State where a reasonable estimate of prescribed fire emissions is desired. Data collection for wildfire emissions estimation will be more difficult because some of the needed information is not currently available in a way that is compatible with emissions estimation requirements. For example, a single wildfire often burns through many different fuel types, but current reporting requirements request the fuel type at the point of ignition. This fuel type may or may not be representative of the majority of acres burned in the wildfire. Also, acres burned in wildland fires may be the area within the fire perimeter rather than the actual acres blackened by fire as is needed for emissions estimation. Similarly, the area reported as burned in prescribed fires is often the area authorized for burning whether or not the entire burn was completed.

State Emission Inventories

High quality Statewide inventories of daily emissions from prescribed fire have been developed by Oregon and Washington since the 1980s (Hardy and others 2001). Eleven other States (Alabama, Alaska, Arizona, California, Colorado, Florida, Idaho, Montana, Nevada, South Carolina, and Utah) estimate annual prescribed fire emissions from records of acreage burned by fuel type and fuel loading at the end of the burning season. Many other States (such as Michigan, New Mexico, and Tennessee) currently have no annual reporting program.

No State has a reporting system for wildland fires that is based on actual, reported data from individual wildland fires events. Any estimate a State may have of wildland fires emissions is based on gross assumptions about fuel loading and consumption, and on an area-burned figure that may systematically overestimate the true value.

Regional Emission Inventories

Several recent regional inventories compiled in support of regional haze program development have shown new approaches to fire emission inventory development.

The Fire Emissions Project (FEP) calculated an emissions inventory for 10 Western States for a current year (1995) using actual reported data, plus two future years (2015 and 2040) using manager projections of fire use. Fourteen vegetative cover types were chosen to characterize the range of species types within the 10-State domain. Within each vegetative cover type, up to three fuel loading categories (high, medium, and low) could be specified by field fire managers. Fuel consumption calculations relied on expert estimates of fuel moisture believed to be most frequently associated with a particular type of burning. Emission factors were assigned based on the vegetative cover type. The FEP inventory was used during the Grand Canyon Visibility Transport Commission (GCVTC) effort to apportion sources of visibility impairment in the Western States.

46

USDA Forest Service Gen. Tech. Rep. RMRS-GTR-42-vol. 5. 2002

The GCVTC also sponsored the development of a wildland fire emissions inventory for the period 1986 through 1992. The GCVTC wildfire inventory included only wildland fires greater than 100 acres in size (capturing approximately 98 percent of the acreage burned). The variability of wildland fire emissions, which ranged from 50,000 tons per year of PM2.5 to more than 550,000 tons per year over the 7 years studied, indicates the difficulty in selecting a single 1 year period that is representative of "typical" fire emissions (GCVTC 1996a).

In 1998, analysts at the Forest Service's Missoula Fire Sciences Laboratory, Rocky Mountain Research Station, used the FEP management strategies with new, additional data to estimate emissions from wildland fires in the Western States (Hardy and others 1998). This inventory of potential emissions used a suite of new or improved spatial data layers, including vegetation/cover type, ownership, fuel and fire characteristics, modeled emissions and heat release rates, and fuels treatment probability distributions. These inventories are included in the Environmental Protection Agency's (EPA) National Emission Inventory (NEI).

Wildland fire frequency and occurrence are highly variable in time and space (fig. 7-1). The impact of wildland fire smoke on Class I area visibility is also expected to be highly variable from year to year with episodic air quality and visibility impact events that are difficult to predict. Seasonal impacts may be many times higher than annual averages.

National Emission Inventories

National emission inventories for prescribed fire have been compiled and reported by several investigators (Chi and others 1979; Peterson and Ward 1992; Ward and others 1976; Yamate and others 1975). Of these, only the Peterson and Ward inventory of particulate matter and air toxic emissions from prescribed fires during 1989 is still useful today, despite the inconsistencies in the information available to compile the emission estimates. The poor data collection and inconsistent or nonexistent reporting systems in use at the time of the 1989 inventory continue today.

Improving Emission Inventories

Significant barriers to compiling better regional inventories include:

- Varying degrees of availability and number of records describing burning activity over multiple States, multiple agencies, ownerships, and Tribes.
- Lack of a national wildland fuel classification system with spatial attributes.
- Limited and inappropriate modeling of fuel consumption and emission characterization for prescribed burning in natural fuels.

Sandberg and others (1999) describe remedies to overcome some of the limitations of data collection and availability. These remedies are intended to guide

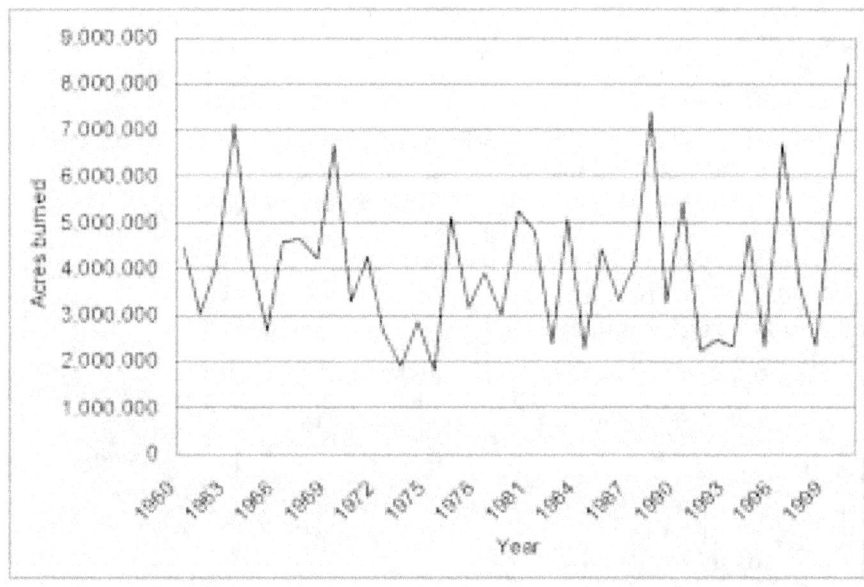

Figure 7-1—Number of acres burned by wildfire between 1960 and 2000 (National Interagency Fire Center 2002).

USDA Forest Service Gen. Tech. Rep. RMRS-GTR-42-vol. 5. 2002

47

future inventory development efforts. Significantly, these remedies include adoption of standardized burn reporting protocols to be used by all agencies, Tribes, and ownerships to report daily emissions for each burn, location of the burn, and many other parameters.

The Fuel Analysis, Smoke Tracking, and Report Access Computer System (FASTRACS) is a sophisticated system developed by the Forest Service and Bureau of Land Management in the Pacific Northwest. FASTRACS tracks all the information needed for accurate estimation of emissions from Federal use of prescribed fire in Washington and Oregon including the ability to track use of emission reduction techniques. As long as field fire managers are doing a reasonable job of reporting the information required by FASTRACS, this system provides excellent emissions calculation capabilities and the best data reporting standards in the country. Currently other landowners, such as State and private, are not using FASTRACS in Washington and Oregon although there is an effort under way to bring them into the system. FASTRACS is also being looked at by other regions and may be adopted or emulated across the country. For more information about FASTRACS, see http://www.fs.fed.us/r6/fastracs/index.htm.

Another data reporting system is under development in California. The Prescribed Fire Information Retrieval System (Cal/PFIRS) is a centralized electronic database that allows all users immediate access to detailed information on burns on a day-to-day basis. Cal/PFIRS does not include the kind of detailed reporting of information that could be used to assess the use of emission reduction techniques but does provide a reasonable estimate of the amount of burning taking place. For more information on Cal/PFIRS, see http://www.arb.ca.gov/smp/progdev/techtool/pfirs.htm.

Research since about 1970 has significantly improved the completeness and accuracy of emission inventory techniques. However, the science is being pressed forward because of new demands for regional scale emission transport information needed to assess the impact of wildland smoke on PM2.5 air quality standards and regional haze. Because of new air regulatory demands, emission inventories, when used in concert with regional models, have become an important means of apportioning fire smoke impacts on air resources.

Air Quality Monitoring

Unlike emission inventories, air quality monitors determine actual pollutant loading in the atmosphere and are therefore the most direct measure of air quality on which air regulatory programs are based. Samples of particulate matter in the atmosphere (PM10 or PM2.5, or both) are also used for source apportionment purposes to identify the origin of the aerosols. Monitoring of smoke from fires, however, presents several unusual technical challenges that affect results. These challenges center on the fact that smoke from fires has several unique characteristics.

Current Monitoring Techniques

The three principal methods of measuring air pollution are samplers, optical instruments, and electrochemical devices. Samplers are most common for long-term monitoring. Data from optical meters and electrochemical devices can be stored in a computer or datalogger on site or transmitted from remote locations to provide real-time information.

Samplers—Samplers collect aerosols on a filter or chemical solution. A simple gravimetric measure of mass concentration may be obtained, or different types of filters or solutions can be used, to help define chemical species and particle sizes. For chemical speciation, filters must be sent to a laboratory for analysis. For this reason, sampling information usually is delayed by days to weeks after the sampling period. Active samplers are the most accurate as they use a pump to pass a known volume of air through the collector. Passive samplers are the least expensive, allowing air to reach the collector by some physical process such as diffusion. Tapered Element Oscillation Microscales (TEOMs) are a special class of samplers that provide a gravimetric measure of mass concentration at the studied site without having to transport filters to a laboratory.

All sampling devices lose some degree of semivolatile fine particulates (Eatough and Pang 1999). Positive and negative organic carbon artifacts are just two of several factors that contribute to variability between different colocated instruments. To minimize this variability, consistent sampling methods are used throughout a sampling network to help recognize such artifacts.

The analytical technique used to quantify carbon concentrations from filters also can cause discrepancies between measurements (Chow 2000). For example, the NIOSH 5040 method (Cassinelli and O'Conner 1994) is a thermal-optical transmittance method of speciating total, organic, elemental, and carbonate (inorganic) carbon being adopted by the EPA's PM2.5 program. This method is a departure from the thermal-optical reflectance method that has been used in the IMPROVE program. Recent comparisons between ambient samples have identified differences as great as 17.5 ± 15 percent (EPA 2000a), which can be significant when monitoring for National Ambient Air Quality Standards (NAAQS) violations.

48

USDA Forest Service Gen. Tech. Rep. RMRS-GTR-42-vol. 5. 2002

Because filters can become overfull, they must be changed regularly and are not suitable for sites close to fires where particulate concentrations are heavy.

Optical Instruments—Optical instruments use a light source to measure the atmosphere's ability to scatter and absorb light. Common devices are photometers, which measure the intensity of light, and transmissometers, which are photometers used to measure the intensity of distant light. Photometers and transmissometers have a direct relation to visual range. Nephelometers measure the scattering function of particles suspended in air. They can be used to determine the visual range, as well as the size of the suspended particles, by changing the wavelength of the light source. Wavelengths of 400 to 550 nm are common for monitoring smoke from biomass fires, while wavelengths of 880 nm are more common for road dust measurements. Because the instruments have increasing application for both long-term and real-time monitoring of smoke, Trent and others (2000) evaluated the accuracy of several different optical instruments by comparing their output to gravimetric samples.

Investigators have found some problems in field reliability and temperature drift among photometers and nephelometers (Trent and others 1999, 2000). While Davies (2002) recommends a general coefficient for relating scattering coefficient to drift smoke from a DataRAM nephelometer, a precise relation between a nephelometer's measured scattering coefficient and particle concentration depends on the wavelength of the instrument and the particle distribution of the medium, which varies by combustion stage and fuel type.

Electrochemical Devices—Electrochemical devices have been used in industrial applications for many years. Their small size and ability to measure criteria pollutants, such as carbon monoxide, make them suitable for personal monitoring or monitoring in extremely remote locations. Thus, they are gaining value for monitoring wildland smoke impacts. For example, Reinhardt and Ottmar (2000) recommend the use of an electrochemical dosimeter for monitoring exposure levels experienced by wildland fire fighters (Reinhardt and Ottmar 2000).

States, Tribes, and local air agencies use a variety of instruments to monitor long-term and real-time smoke impacts for both NAAQS and visibility to suit their local interests and regulatory needs. The Interagency Monitoring of Protected Visual Environments (IMPROVE) program is one of few nationally coordinated monitoring projects.

IMPROVE was established in 1985 in response to the 1977 amendment of the Clean Air Act requiring monitoring of visibility-related parameters in Class I areas throughout the country (fig. 7-2). The IMPROVE network uses a combination of speciation filters on active samplers to measure physical properties of atmospheric particles (PM2.5 and PM10) that are related to visibility. Many sites also include transmissometers and nephelometers optical devices. Also, cameras are used document the appearance of scenic vistas. Because the samplers collect for 24 hours every 3 days, their information is used for determining long-term trends in visibility. The optical and camera devices can monitor more frequently and can help define short-term or near real-time changes in visibility impact

Source Apportionment

Most air monitoring programs are designed to measure particulate mass loading to provide data for PM10 and PM2.5 NAAQS and visibility. Because these sizes of particles can come from many sources, they are not useful for apportioning to one source or another. While the IMPROVE program provides speciated aerosol data that are helpful in source attribution analysis, the averaging periods of samples and sparse location of sites make IMPROVE measurements difficult to use for source attribution without supplemental measurements or modeling tools.

Wotawa and Trainer (2000) found that 74 percent of the variance in the average afternoon carbon monoxide levels could not be attributed to anthropogenic sources during the 1995 Southern Oxidant Study (Chameides and Cowling 1995). Analysis of weather patterns indicated that transport of wildland fire smoke from Canada could explain the elevated carbon monoxide levels. Also, they discovered a statistically significant relationship between the elevated carbon monoxide and ground-level ozone concentrations.

Characterization of organic carbon compounds found within the organic carbon fraction of fine particulate matter coupled with inclusion of gaseous volatile organic compounds (VOCs) holds substantial promise in advancing the science of source apportionment (Watson 1997). The key to the use of chemical mass balance methods is the acquisition of accurate data describing the chemical composition of both particulate matter and VOCs in the ambient air and in emissions from specific sources. Several organic compounds unique to wood smoke have been identified including retene, levoglucosan, thermally altered resin, and polycyclic aromatic hydrocarbons (PAH) compounds. These compounds are present in appreciable amounts and can be used as signatures for source apportionment if special precautions are taken during sampling to minimize losses (Standley and Simoneit 1987). Inclusion of these aerosol and VOC components in the speciation

USDA Forest Service Gen. Tech. Rep. RMRS-GTR-42-vol. 5. 2002

49

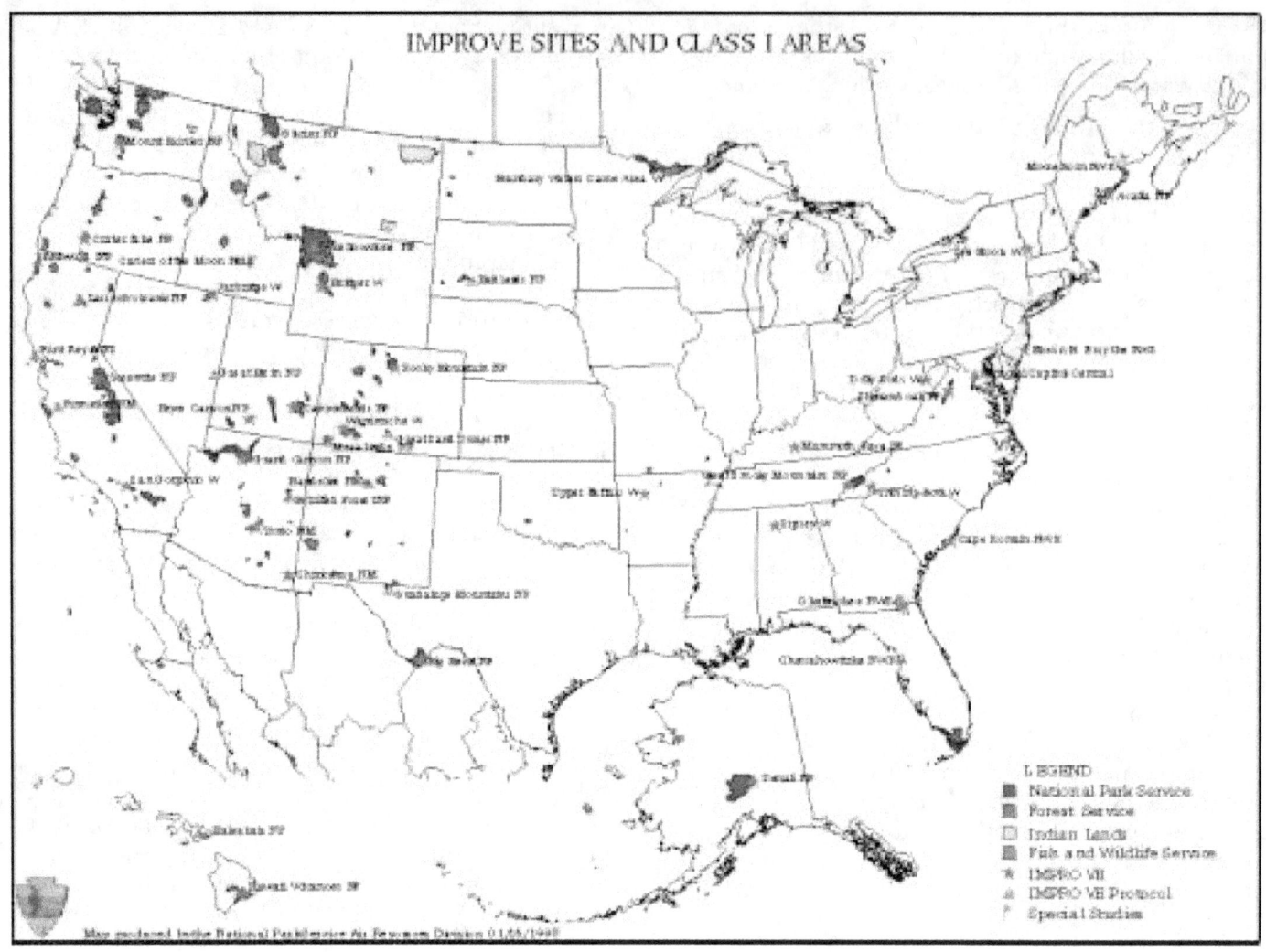

Figure 7-2—IMPROVE monitoring network in 1999 (http://vista.cira.colostate.edu/improve/Overview/IMPROVEProgram.htm).

analysis appears worthwhile but would increase monitoring and sample analysis costs.

Source Apportionment Methods

Apportionment of particulate matter mass to the respective contributing sources is done through both mechanistic models (dispersion models) and receptor-oriented techniques that are based on the characteristics of the particles collected at the receptor. The best approach is through the use of both techniques, applied independently, to develop a "weight of evidence" assessment of source contributions of smoke from fire. A third approach is through the use of visual and photographic systems that can document visibility conditions over time or track a plume from its source to the point of impact within a Class I area.

Receptor-Oriented Approaches

Receptor-oriented approaches range from simple signature applications to complex data analysis techniques that are based on the spatial, temporal, and chemical constituents ("fingerprint") of various sources.

Simple signature applications for smoke from fire are based on chemically distinct emissions from fire. For example, methyl chloride (CH_3Cl) is a gas emitted during wood combustion that has been used in this manner to identify impacts of both residential woodstove smoke and smoke from prescribed fires (Khalil and others 1983).

Speciated Rollback Model —The speciated rollback model (NRC 1993) is a simple hybrid model that uses aerosol data collected at the receptor with emission inventories to estimate source impacts. It is a

50

USDA Forest Service Gen. Tech. Rep. RMRS-GTR-42-vol. 5. 2002

spatially averaged model that disaggregates major particle components into chemically distinct groups that are contributed by different types of sources. A linear rollback model is based on the assumption that ambient concentrations (C) above background (C_b) are directly proportional to total emissions in the region of interest (E):

$$C - C_b = kE \qquad (1)$$

The proportionality constant, k, is determined over a historical time period when both concentrations C and C_b as well as regional emissions E are known. Once k is determined, new concentration estimates can be derived for other emission levels of interest assuming that meteorological conditions are constant over the same averaging time. Because the anthropogenic components in the particle mass consist almost entirely of sulfates, nitrates, organic carbon, elemental carbon, and crustal material, a maximum contribution from fire can be made based on the assumption that all of the organic carbon or elemental carbon is from primary fire emissions. Various complexities can be added to this model; components can be disaggregated by particle-size fraction (coarse versus fine particles) as well as by chemical composition. Additional distinctions can be made between primary and secondary particles, and nonlinear transformation processes can be approximated to account for atmospheric reactions.

Simple proportional speciated rollback models require data on the chemical composition of airborne particles, knowledge or assumptions regarding secondary particle components, an emission inventory for the important source categories for each particle component and each gaseous precursor, and knowledge or assumptions regarding background concentrations for each component of the aerosol and each gaseous precursor.

The speciated rollback model was applied by the NRC Committee on Haze in National Parks and Wilderness Areas to apportion regional haze in the three large regions of the country (East, Southwest, and Pacific Northwest) by including extinction coefficients to the estimated mass concentrations (NRC 1993). The percentage of anthropogenic light extinction apportioned to forest management burning was estimated at 11 percent in the Northwestern United States on an annual basis assuming that about one-third of the measured organic carbon is of natural origin. The 1985 National Acid Precipitation Assessment Program (NAPAP) inventory was used in this analysis, which also assumed that the elemental carbon and organic carbon fractions of the PM2.5 emissions for forest management burning were 6 percent and 60 percent, respectively.

The limitations of the speciated rollback model are several:

- Deviations from the assumption of spatially homogeneous emissions are likely to occur when air quality is most critical at a single receptor where a single emission source can have an inordinate impact.
- Secondary particle formation is assumed to be linear to changes in precursor emissions.
- Meteorological conditions do not change from year to year.
- Emission inventory errors have a direct, proportional effect on the model estimates.

The model can be applied to any temporal concentration such as annual average, worst 20th percentile, or worst daily average scenarios in any region that meets the constraint on the spatial distribution of emission changes. It is straightforward, necessary input data are available, and the model assumptions are easily understood. It makes use of chemical speciation data collected from the IMPROVE network but cannot apportion contributions made from source classes not included in the inventory.

Chemical Mass Balance Model—The chemical mass balance model, CMB7 (Watson 1997; Watson and others 1990), infers source contributions based on speciated aerosol samples collected at a monitoring site. Chemical elements and compounds in ambient aerosol are "matched" to speciated source emission profiles "fingerprints" by using least-squares, linear regression techniques to apportion the aerosol mass. CMB7 has been widely used within the regulatory community to identify and quantify the sources of particles emitted directly to the atmosphere. The model is based on the relationship between characteristics of the airborne particle (c_i), the summation of the product of the ambient mass concentration contributed by all sources (S_j), and the fraction of the characteristic component in the source's fingerprint (f_{ij}).

$$c_i = \sum_j S_j f_{ij} \qquad (2)$$

Given detailed information about the chemical speciation of the ambient aerosol and similar information about all of the emission sources impacting the receptor, the CMB7 model can apportion the aerosol mass among the sources if certain assumptions are met.

To minimize error, there must be more aerosol components than sources to be included in the least-squares linear regression fit. If there are more components measured than sources, then the comparison of model-estimated concentrations of these additional components provides a valuable internal check on model consistency.

USDA Forest Service Gen. Tech. Rep. RMRS-GTR-42-vol. 5. 2002

51

The chemical components in the source "fingerprint" must be conserved and not altered during atmospheric transport — a rather large limitation.

Model resolution is typically limited to five or six source types, and separation of two sources with similar emission profiles (for example, prescribed burning and residential woodstove smoke) is difficult if both sources are active at the same time.

Systematic error analysis procedures have been developed for the CMB7 model, and the results have been published in model validation studies (NRC 1993). However, the model cannot apportion secondary aerosols (sulfate and nitrate); it is limited in its ability to apportion all of the mass to specific sources.

The ability of the model to apportion smoke from fire depends on several factors:

- The presence or absence of smoke from other forms of vegetative burning (woodstoves, agricultural burning, open burning, and others).
- The magnitude of the smoke impact at the receptor (for example, well-dispersed smoke that contributes small amounts of aerosol mass is more difficult to distinguish).
- The uncertainty in both the ambient aerosol and the source "fingerprint" components that the model most heavily weights in the regression analysis, typically organic carbon, potassium, and elemental carbon. The greater the uncertainty of these measurements, the less "fitting pressure" they have in influencing the regression solution.
- Inclusion of multiple aerosol components that are as nearly unique to smoke from fires (endemic signatures) as possible. These include organic compounds such as retene and levoglucosan, as well as gaseous signature such as carbon monoxide and methyl chloride. The more the source profile distinguishes prescribed or wildland fire smoke from other sources, the more accurate the source apportionment is likely to be.

Factor Analysis and Multiple Linear Regression

When many ambient samples are available, linear regression and factor analysis techniques can be applied to the dataset to obtain empirical insights into the origin of the particles. Factor analysis is based on the assumption that chemical components of the aerosol that covary are emitted from a common source. Cluster patterns can then be matched to the source profiles of known sources to identify the degree of covariance associated with a specific source category.

Source profiles can be recovered from the ambient data by using two special forms of factor analysis (VARIMAX rotation) or, when the profiles are approximately known, target transformation factor analysis (Hopke 1985). Factor analysis can therefore serve to refine the source profile information used in chemical mass balance analysis. In the context of wildfire smoke apportionment, investigators have historically looked for a high degree of covariance between organic carbon, elemental carbon, and potassium (total, water soluble, and/or nonsoil potassium) as the cluster components that signal particles emitted from vegetative burning of all kinds. Unfortunately, these components of the aerosol are not necessarily unique to smoke from vegetative burning.

Linear regression analysis is a well-established statistical procedure for estimating unknown coefficients in linear relationships where a large dataset of observations of both the dependent and independent variables are present. In the terminology of regression analysis, c in equation (2) is the variable, Sj is the independent variable, and fj the regression coefficients. In practice, the independent variable is taken to be proportional to source strength rather than the source strength themselves. Multiple linear regression has been widely used to apportion total particle mass, the most common approach being use of signature concentrations taken directly as the independent variable, fj. Significantly, gaseous pollutant data can be included in the regression to increase the model's ability to resolve sources. Although carbon monoxide would greatly enhance the success of the model as it is emitted by wildfires in large quantities and is stable in the atmosphere, carbon monoxide is not routinely included in nonurban monitoring programs.

Regression analysis has been used successfully to apportion the total carbon portion of the aerosol mass between wood smoke, vehicle exhaust, and other sources by using nonsoil potassium . The regression-derived estimates were then validated by ^{14}C isotope analysis, which is a direct indicator of "contemporary" versus fossil fuel carbon sources. The ^{14}C measurements nicely confirmed the source apportionment results by regression analysis (r=0.88) (NRC 1993).

Summary

Receptor-oriented methods of particle mass source apportionment have proven successful in a large number of urban studies worldwide. A number of these studies have attempted to apportion wildfire smoke on the basis of a set of aerosol and source emission trace elements and compounds. The experimental design of these studies has limited the ability of receptor models to resolve wildfire smoke from other sources. With improvements in speciation of the organic carbon component of the aerosol, and inclusion of carbon monoxide, methyl chloride, and other endemic signatures, the ability of these techniques to resolve sources and minimize uncertainties will increase. Sensitivity studies are needed to determine which additional components beyond the standard array of trace elements, ions, and carbon fractions would be most beneficial to include in future monitoring programs.

Mechanistic Models

As noted in chapter 6, multiple dispersion models have been used to estimate air quality impacts of single or multiple fires at local and regional scales. Eulerian regional-scale models have been principally used for source apportionment application both to estimate contributions to particulate air quality and regional haze. The suitability of such models for apportionment applications largely depends on the completeness and accuracy of the emission inventory inputs used by the model. Unfortunately, few field validations are available.

USDA Forest Service Gen. Tech. Rep. RMRS-GTR-42-vol. 5. 2002

53

Notes

_____

_____

_____

_____

_____

_____

Chapter 8: Consequences of Fire on Air Quality

The potential impacts of fire-induced degradation of air quality on public health and welfare range from occupational exposure of smoke on firefighters to broader economic and social impacts and highway safety.

Health Effects _____

National Review of Health Effects

In 1996, the Environmental Protection Agency (EPA) conducted an extensive review of the science relating human health effects to particulate matter (PM), the principal pollutant of concern from fires (EPA 1996). The review found that (1) epidemiological studies suggest a variety of health effects at concentrations found in several U.S. cities and (2) ambient particles of greatest concern to health were those smaller than 10 micrometers in diameter. Results of efforts to trace the physiological and pathological responses of the body to PM are unclear, and demonstration of possible mechanisms linking ambient PM to mortality and morbidity are derived from hypotheses in animal and human studies. It is known, however, that PM produces physiological and pathological effects by a variety of mechanisms, including:

- Increased airflow obstruction by PM-induced narrowing of airways.
- Impaired clearance of lung pathways caused by hypersecretion of mucus caused by PM exposure.
- Lung responses to PM exposure including hypoxia, broncho-constriction, apnea, impaired diffusion and production of inflammatory mediators.
- Changes in the epithelial lining of the alveolar capillary membrane that increase the diffusion distances across the respiratory membrane, thereby reducing the effectiveness of blood gas exchange.
- Inflammatory responses that cause increased susceptibility to asthma, chronic obstructive pulmonary disease (COPD) and infections.

Recent information also suggests that several subgroups within the population are more sensitive to PM than others. Children are more likely to have decreased pulmonary function, while increased mortality has been reported in the elderly and in individuals with cardiopulmonary disease. Asthmatics are especially susceptible to PM exposure. In addition, coarse (2.5 to 10μm) particles from road dust or windblown soil were found to have less toxicity than fine particles (less than 2.5μm) that include acid aerosols, diesel

USDA Forest Service Gen. Tech. Rep. RMRS-GTR-42-vol. 5. 2002

55

emissions, smoke from fires, and potentially carcinogenic PAH compounds.

Occupational Exposure to Wildland Fire Smoke

Wildland firefighters and fire managers have long been aware that smoke exposure occurs during their work (Reinhardt and Ottmar 1997; Sharkey 1997). Although the long-term health effects from occupational smoke exposure remain unknown, the evidence to date suggests that brief, intense smoke exposures can easily exceed short-term exposure limits in peak exposure situations such as direct attack and holding firelines downwind of an active wildfire or prescribed burn. Shift-average exposure only occasionally exceeds recommended instantaneous exposure limits set by the American Conference of Governmental Industrial Hygienists (ACGIH), and rarely do they exceed Occupational Safety and Health Administration (OSHA) time weighted average (TWA) limits (fig. 8-1) (Reinhardt and Ottmar 2000; Reinhardt and others 2000). Overexposure increases to 10 percent of the time if the exposure limits are adjusted for unique aspects of the fire management workplace; these aspects include hard breathing, extended hours, and high elevations, all factors which intensify the effects of many of the health hazards of smoke (Betchley and others 1995; Materna and others 1992; Reinhardt and Ottmar 2000; Reinhardt and others 2000). It could be argued that few firefighters spend a working lifetime in the fire profession, and thus they should be exempt from occupational standards

Figure 8-1—Firefighters being monitored for smoke exposure. Monitoring equipment seen includes a red backpack that collects gas samples from the breathing zone of the firefighters and a white-colored particulate matter filter sampler attached to the chest. (Photo by Roger Ottmar)

that are set to protect workers over their careers. But this argument is irrelevant for irritants and fast-acting health effects such as eye and respiratory irritation, headache, nausea, and angina. An exposure standard specifically for wildland firefighters and appropriate respiratory protection may need to be developed (Reinhardt and Ottmar 2000).

In spite of the studies that have been done, major data gaps remain:

- In the area of health hazards, not enough evidence is available to defend the commonly cited "inert" classification of total and respirable particulate in dust and smoke; there is little knowledge of the occurrence of crystalline silica in dust at fires; and there is incomplete characterization of aldehydes and other respiratory irritants present in smoke (Reinhardt and Ottmar 1997, 2000).
- The differences in smoke exposure between large and small wildland fires have not been characterized in spite of the fact that one or two crews extinguish the vast majority of wildfires (Reinhardt and Ottmar 2000).
- The long-term health experience of wildland firefighters is unknown, although anecdotal reports and the biological plausibility of cumulative health effects indicate a potentially greater incidence of disease and death than in the general population of workers (Booze and Reinhardt, in press; Sharkey 1997).

Although data gaps remain, enough information has been gathered to chart a course to alleviate many of the overexposures. Respiratory protection is available for irritants such as aldehydes and particulate matter but not for carbon monoxide. Respirators can be heavy, hot, and impede the speed of work, but some new models are light, simple and could be worn only when needed (Beason and others 1996; Rothwell and Sharkey 1995). The entire costly process of medical evaluations, fitness testing, maintenance, and training must be employed if respirators are to be used. But there are immediate benefits to reducing respiratory irritant exposure. Small electrochemical dosimeters can provide instant warnings about carbon monoxide levels in a smoky situation, and fire crew members equipped with respirators and carbon monoxide monitors have all the protection necessary to stay and accomplish objectives safely and withdraw when the carbon monoxide levels become the limiting factors (Reinhardt and others 1999). In the future, a respirator for use during wildland fires may be developed that offers warning and protection against carbon monoxide as well. Although some work has been done in this area, we need more significant development. Smoke exposure is a hazard only a small portion of the time and is manageable because the situation where it occurs can be predicted. A long-term program to manage smoke exposure at wildland fires could include (1) hazard awareness training, (2) implementation of practices to reduce smoke exposure such as rotating crews and providing clean air sites, (3) routine carbon monoxide monitoring with electronic dosimeters, (4) improved recordkeeping on accident reports to include separation of smoke related illness among fireline workers and fire camp personnel, and (5) improved nutritional and health habits. Fire management practices such as crew rotation, awareness training, and carbon monoxide monitoring can mitigate the hazard and allow firefighters to focus on the job of fire management, lessening the distraction, discomfort, and health impacts of smoke exposure (Reinhardt and Ottmar 2000).

Research Issues

A number of wildland fire health effect research issues flow from the EPA staff report (Clean Air Scientific Advisory Committee 1995) and occupational health exposure studies.

Research into the health effects of particulate matter is largely based on epidemiological studies conducted over long periods in urban centers with high hospital admittance or large air quality databases, or both. Consequently, inadequate information is available that relates short-term, acute smoke exposure (such as would be experienced by a visitor to a National Park or to a community near a wildfire) to human health effects. As a result, little or no specific guidance is available to wildland fire managers, air quality regulators, or public health officials who need to responsibly judge the public health risks of exposure to extremely high smoke concentrations. This gap in knowledge was clearly evident during the 1988 Yellowstone fires and later wildfire events when quick decisions had to be made on how best to protect public health in communities near major wildfires (WESTAR 1995). The best available guidelines are those published by EPA (1999) for assessing the risk to health from air pollution (table 8-1). These guidelines may or may not reflect the specific hazards of pollutants from fires, which will have a different chemical composition.

Table 8-1—Pollutant-specific breakpoints for the air quality index (AQI) and accompanying health effects statements (adapted from EPA 1999).

Category	PM2.5 (24-hour)		PM10 (24-hour)	
	Concentration breakpoints	Health effects statements	Concentration breakpoints	Health effects statement
	$\mu g/m^3$		$\mu g/m^3$	
Good	0.0-15.4	None	0-54	None
Moderate	15.5-40.4	None	55-154	None
Unhealthy for sensitive groups	40.5-65.4	Increasing likelihood of respiratory symptoms in sensitive individuals, aggravation of heart or lung disease and premature mortality of persons with cardiopulmonary disease and the elderly.	155-254	Increasing likelihood of respiratory symptoms and aggravation of lung disease, such as asthma.
Unhealthy	65.5-150.4	Increased aggravation of heart or lung disease and premature mortality in persons with cardiopulmonary disease and the elderly; increased respiratory effects in the general population.	255-354	Increased respiratory symptoms and aggravation of lung disease, such as asthma; possible respiratory effects in general population.
Very unhealthy	150.5-250.4	Significant aggravation of heart or lung disease and premature mortality in persons with cardiopulmonary disease and the elderly; significant increase in respiratory effects in general population.	355-424	Significant increase in respiratory symptoms and aggravation of lung disease, such as asthma; increasing likelihood of respiratory effects in general population.
Hazardous	250.5-500.4	Serious aggravation of heart or lung disease and premature mortality in persons with cardiopulmonary disease and the elderly; serious risk of respiratory effects in general population.	425-604	Serious risk of respiratory symptoms and aggravation of lung disease, such as asthma; respiratory effects likely in the general population.

USDA Forest Service Gen. Tech. Rep. RMRS-GTR-42-vol. 5. 2002

57

The long-term health effects of smoke exposure to wildland firefighters are unknown in spite of anecdotal evidence that indicates the possibility of a greater incidence of cardiopulmonary disease and death than in the general population. Although carbon monoxide monitoring and respiratory protection can mitigate the hazard, personal protection equipment is still needed that allows firefighters to work effectively without discomfort or distraction (Reinhardt 2000).

Welfare Effects

Air quality-related effects of smoke include the soiling of materials, public nuisance, and visibility loss. Because these and other consequences of smoke have come increasingly into conflict with the public's interest in clean air, an understanding of these effects is important to fire managers.

Soiling of Materials

The deposition of smoke particles on the surface of buildings, automobiles, clothing, and other objects reduces aesthetic appeal and damages a variety of objects and building structures (Baedecker and others 1991). Studies of the effect of aerodynamic particle size on soiling have concluded that coarse particles (2.5 to 10μm) initially contribute more to soiling of both horizontal and vertical surfaces than do fine particles (less than 2.5μm), but that coarse particles are more easily removed by rainfall (Haynie and Lemmons 1990). Smoke from fires is largely within the fine mode, although ash fallout in the near vicinity of a fire is often also a concern. Smoke may also discolor artificial surfaces such as building bricks or stucco, requiring cleaning or repainting. Increasing the frequency of cleaning, washing, or repainting soiled surfaces becomes an economic burden and can reduce the life usefulness of the soiled material (Maler and Wyzga 1976).

Soiling from smoke also changes the reflectance of opaque materials and reduces light transmission through windows and other transparent materials (Beloin and Haynie 1975).

When fine smoke particles (less than 2.5μm) infiltrate indoor environments, soiling of fabrics, painted interior walls, and works of art may occur. Curtains may require more frequent washing because of soiling or may deteriorate along folds in the fabric after being weakened by particle exposure (Yocom and Upham 1977). As in the case of corrosion damage from acidified particles, these same particles accelerate damage to painted surfaces (Cowling and Roberts 1954). Studies of the soiling of works of art at a museum in southern California concluded that a significant fraction of the dark-colored fine mode elemental carbon and soil dust originated from outdoor sources (Ligocki and others 1993). Smoke from fires is one source of elemental carbon.

Public Nuisance and Visibility Loss

Nuisance smoke is the amount of smoke in the ambient air that interferes with a right or privilege common to members of the public, including the use or enjoyment of public or private resources (EPA 1990). The abatement of nuisance smoke is one of the most important objectives of successful smoke management (Shelby and Speaker 1990). Public complaints about nuisance smoke are linked to loss of visibility, odors, and ash fallout that soils buildings, cars, laundry, and other objects. Acrolein (and possibly formaldehyde) in smoke at distances of 1 mile from the fireline are likely to cause eye and nose irritation, exacerbating public nuisance conditions (Sandberg and Dost 1990).

Perhaps the most significant nuisance effect of smoke from fire is local visibility reduction in areas impacted by the plume. While visibility loss within Class I areas is subject to regulation under the Clean Air Act, smoke plume-related visibility degradation in urban and rural communities is not. Nuisance is usually regulated under State and local laws and is frequently based on public complaint or, when highway safety is compromised, the risk of litigation (Eshee 1995). The courts have also ruled that the taking of private property by interfering with its use and enjoyment caused by smoke (and without just compensation) is in violation of Federal Constitutional provisions under the Fifth Amendment. The trespass of smoke may diminish the value of the property, resulting in losses to the owner (Iowa Supreme Court 1998).

Because the public links visibility loss with concerns about the health implications of breathing smoke, smoke management programs have been under increasing pressure to minimize emissions and reduce smoke impacts to the greatest degree possible (Core 1989). Visibility reduction is used as a measure of smoke intrusions in several smoke management plans. The State of Oregon program operational guidance defines a "moderately" intense intrusion as a reduction of from 4.6 to 11.4 miles from a background visibility of more than 50 miles (Oregon Department of Forestry 1992). The State of Washington smoke intrusion reporting system uses a "slightly visible," "noticeable impact on visibility" or "excessive impact on visibility" to define light, medium, and heavy intrusions (Washington Department of Natural Resources 1993). The State of New Mexico program requires that visibility impacts of smoke be considered in development of the unit's burn prescription (New Mexico Environmental Improvement Board 1995).

Economic and Social Consequences

The economic consequences of smoke are principally in the areas of soiling-related losses and costs related to reduced visibility.

Soiling-Related Economic Losses

Economic costs associated with materials damage and soiling caused by airborne particles include reduction in the useful life of the damaged materials and the decreased utility of the object. Losses caused by the need for more frequent maintenance and cleaning are also significant. Amenity losses occur when the increased cleaning or repair of materials results in inconvenience or delays, many of which are difficult to quantify (Maler and Wyzga 1976).

Within the United States, however, the soiling of buildings constitutes the largest category of surface areas at risk to pollution damage (Lipfert and Daum 1992). Soiling on painted surfaces on residential buildings, resulting in a need to repaint exterior walls, has caused damage approaching $1 billion per year (Haynie and others 1990).

Willingness-to-pay estimates developed using the contingent valuation method found that households were willing to pay $2.70 per $\mu g/m^3$ charge in particle pollution to avoid soiling effects (McClelland and others 1991). No estimates are available for costs specifically associated with smoke from fires.

Visibility-Related Costs

The importance of clean, clear air within the wildlands and National Parks of this nation is hard to overemphasize. People go to these special places to enjoy scenery, the color of the landscapes, and clarity of the vistas. At Grand Canyon, 82 percent of 638 respondents rated "clean, clear air" as very important or extremely important to their recreational experience (Ross 1988). Three National Park Service (NPS) studies determined that air quality conditions affect the amount of time and money visitors are willing to spend at NPS units (Brookshire and others 1976; MacFarland and others 1983; Schulze and others 1983). These studies found estimated onsite use values for the prevention or elimination of plumes that ranged from about $3 to $6 (1989 dollars) per day per visitor party at the park. Based on these results, the implied preservation value for preventing a visible plume most days (the exact frequency was not specified) at the Grand Canyon was estimated at about $5.7 billion each year when applied to the total U.S. population (EPA 1996). Other investigators have suggested that these estimates are overstated by a factor of 2 or 3 (Chestnut and Rowe 1990).

In the studies noted above, park visitors generally responded that they would be willing to spend more time and money if visibility conditions were better and, conversely, less if visibility conditions were worse (Ross 1988). The average amount of time visitors were willing to spend traveling to a vista for every unit change in visibility (.01 km^{-1} extinction coefficient) was between 15 minutes and 4 hours. These results provide evidence that changes in visual air quality can be expected to affect visitor enjoyment and satisfaction with park visits.

Even given the limitations and uncertainties of contingent valuation surveys, economic values related to visibility degradation are clearly likely to be substantial.

Public Perception of Haze—Perceived visual air quality (PVAQ) has been used as a measure of the public's acceptance of haze conditions (Middleton and others 1983). Subjects were asked to judge the visual air quality in several photos depicting vistas under different haze conditions using a scale of 1 to 10, 1 being the worst and 10 being the best. These 1 to 10 scales reflect people's perceptions and judgments concerning visibility conditions. By matching particulate air quality conditions that occurred at the time of the photographs, researchers have been able to develop a relationship between PVAQ and particulate matter concentrations (Middleton and others 1985). Even small increases in particulate concentrations in the atmosphere result in dramatic decreases in PVAQ. Because of the light scattering efficiency of smoke, this relationship is especially applicable to fire emissions.

Cultural Consequences of Visibility Loss—"National parks and wilderness areas are among our nation's greatest treasures. Ranging from inviting coastal beaches and beautiful shorelines to colorful deserts and dramatic canyons to towering mountains and spectacular glaciers, these regions inspire us as individuals and as a nation" (NRC 1993). With these words, the National Research Council (NRC) noted the importance of preserving the scenic vistas of the nation. Congress, in recognition of the scenic values of the nation, adopted the Clean Air Act Amendments of 1977, which established a national visibility protection program. The GCVTC was later established in the 1990 amendments to the act to address visibility impairment issues relevant to the region surrounding Grand Canyon National Park. Following 4 years of study, the GCVTC concluded that smoke from wildland fires is likely to have the single greatest impact on visibility in Class I areas of the Colorado Plateau through the year 2040 (GCVTC 1996c). While difficult to quantify, there is consensus that visibility loss associated with smoke from wildland fire and other sources has important cultural consequences on the nation.

USDA Forest Service Gen. Tech. Rep. RMRS-GTR-42-vol. 5. 2002

59

Highway Safety

Smoke can cause highway safety problems when it impedes a driver's ability to see the roadway (fig. 8-2) and can result in loss of life and in property damage at smoke levels that are far below NAAQS. This section focuses on highway safety issues in the Southeastern United States because this is where the foremost forestry-related air quality problem has been in the past. We also describe tools being developed to aid the land manager in avoiding highway safety problems.

Although smoke at times can become a problem anywhere in the country, it is in the Southern States, from Virginia to Texas and from the Ohio River southward, where highway safety is most at risk from prescribed fire smoke, principally because of the amount of burning done in the South and the proximity of wildlands to population centers. Roughly 4 million acres of Southern forests are treated with prescribed fire each year (after Wade and Lunsford 1988). This area is by far the largest acreage subjected to prescribed fire in the country. Prescribed fire treatment intervals, especially in Southern pine (in an area extending roughly from Virginia to Texas), is every 3 to 5 years. These forests are intermixed with homes, small towns, and scattered villages within an enormous wildland/urban interface. During the daytime, smoke becomes a problem when it drifts into these areas of human habitation. At night, smoke can become entrapped near the ground and, in combination with fog, creates visibility reductions that cause roadway accidents. The potential exists for frequent and severe smoke intrusions onto the public roads and highways from both prescribed and wildland fires.

Magnitude of the Problem

Smoke and smoke/fog obstructions of visibility on Southeastern United States highways cause numerous accidents with loss of life and personal injuries every year. Several attempts to compile records of smoke-implicated highway accidents have been made. For the 10 years from 1979 through 1988, Mobley (1989) reported 28 fatalities, over 60 serious injuries, numerous minor injuries, and millions of dollars in lawsuits. During 2000, smoke from wildfires drifting across Interstate 10 caused at least 10 fatalities, five in Florida and five in Mississippi.

As the population growth in the South continues, more people will likely be adversely impacted by smoke on the highways. Unless methods are found to adequately protect public safety on the highways, there exists the prospect that increasingly restrictive regulations will curtail the use of prescribed fire or that fire as a management tool may be altogether prohibited.

Figure 8-2—Smoke can cause highway safety problems when it impedes a driver's ability to safely see the roadway. (Photo by Jim Brenner)

Measures to Improve Highway Safety

Several approaches are being taken to reduce the uncertainty of predicting smoke movement over roadways:

High-resolution weather prediction models promise to provide increased accuracy in predictions of wind speeds and directions and mixing heights at time and spatial scales useful for land managers. The Florida Division of Forestry (FDOF) is a leader in the use of high resolution modeling for forestry applications in the South (Brenner and others 2001). Because much of Florida is located within 20 miles of a coastline, accurate predictions of sea/land breezes and associated changes in temperature, wind direction, atmospheric stability, and mixing height are critical to the success of the FDOF. High-resolution modeling consortia are also being established by the USDA Forest Service to serve clients with interests as diverse as fire weather, air quality, ecology, and meteorology. These centers involve scientists in development of new products and in technology transfer to bring the products to consortia members.

Several smoke models are in operation or are being developed to predict smoke movement over Southern landscapes. VSMOKE (Lavdas 1996), a Gaussian plume model that assumes level terrain and unchanging winds, predicts smoke movement and concentration during daytime. VSMOKE has been made part of the FDOF fire and smoke prediction system. It is a screening model that aids land managers in assessing where smoke might impact sensitive targets as part of planning for prescribed burns. PB-Piedmont (Achtemeier 2001) is a wind and smoke model designed to simulate smoke movement near the ground under entrapment conditions at night. The smoke plume is simulated as an ensemble of particles that are transported by local winds over complex terrain characteristic of the shallow (30 to 50 m) interlocking ridge/valley systems typical of the Piedmont of the South. Two sister models are planned — one that will simulate near-ground smoke movement near coastal areas influenced by sea/land circulations, and the other for the Appalachian Mountains.

Climate Change

Globally, fires are a significant contributor of carbon dioxide and other greenhouse gases in the atmosphere. Fires are also an important mechanism in the redistribution of ecosystems in response to climate stress, which in turn affects the atmosphere-biosphere carbon balance. Currently, there is no policy mandate, nor widely accepted methodology for managing fires, for the conservation of terrestrial carbon pools or mitigation of greenhouse gas emissions. However, we may expect carbon accounting and perhaps conservation to become a part of fire and air resource management if and when global agreements are made to address biomass burning and resultant greenhouse gas emissions.

USDA Forest Service Gen. Tech. Rep. RMRS-GTR-42-vol. 5. 2002

61

Notes

_____

_____

_____

_____

_____

_____

Chapter 9: Recommendations for Future Research and Development

Managing smoke and air quality impacts from fires requires an increasing base of knowledge obtained through research and the development of information systems. Fire and air resource managers have had the responsibility since the 1960s to mitigate direct intrusions of smoke into areas where it presents a health or safety hazard, or where it is simply objectionable to an affected population. In more recent years, that responsibility has broadened because of an increase in the use of fire, more people in the wildland/urban interface, tightening of regulatory standards, and decreasing public tolerance for air pollution. More States require smoke management plans, and the plans are increasingly complex due to increased coverage and greater requirements for notification, modeling, monitoring, and recordkeeping.

Established Research Framework _____

There is ample strategic analysis and workshop output to guide research. The most comprehensive and up-to-date recommendations for research and development are found in *National Strategic Plan: Modeling and Data Systems for Wildland Fire and Air*

Quality (Sandberg and others 1999). Workshop sessions, internal discussion, and review comments were compiled into more than 200 proposals from which 46 priority projects were selected that support the nine summary recommendations outlined here.

Recommendation 1: Fuels and fire characteristics—An ability to estimate emissions from all types of fires over the wide variation in fuels in the contiguous United States and Alaska is needed. Expanded models and fuel characteristics data are needed to fill this gap.

Recommendation 2: Emissions modeling systems—Current models to estimate emissions are inadequate in coverage and incomplete in scope. Emissions production models need to be expanded to include all fire and fuel types as well as linked to fire behavior and air quality models in a geographically resolved data system.

Recommendation 3: Transport, dispersion, and secondary pollutant formation—Air quality and land management planners lack spatially explicit planning and real-time systems for assessing air quality impacts. A geographic information system (GIS) based

USDA Forest Service Gen. Tech. Rep. RMRS-GTR-42-vol. 5. 2002

63

system linked to emissions production, meteorological, and dispersion models is needed.

Recommendation 4: Air quality impact assessment—Better wildland and prescribed fire information is needed to compile emissions inventories, for regional haze analysis and for determination of compliance with National Ambient Air Quality Standards (NAAQS).

Recommendation 5: Emissions tradeoffs and determination of "natural" visibility background assessments—No policy-driven or scientific definition of "natural" background visibility exists for regional haze assessments. The tradeoffs between wildfire and prescribed fire emissions are also not known. To address these issues, the policy community needs to decide what types of fires contribute to natural impairment after which a scientific assessment could be done and tradeoffs evaluated.

Recommendation 6: Impact and risk assessment of emissions from fire—A comprehensive assessment of smoke exposure of prescribed and wildland firefighters and the public at current levels of fire activity should be done to provide a baseline for future risk assessments. Exposures should be periodically reassessed to evaluate increased risks from future increases in fire emissions.

Recommendation 7: Monitoring guidelines and protocols—Guidelines are needed on how best to monitor source strength, air quality, visibility, and nuisance impacts from fires to support consistent and quantitative evaluation of air impacts.

Recommendation 8: National fire and air quality information database—A readily accessible source of information on past, current, and predicted future fire activity levels, emissions production, and air quality impacts from fires does not exist. Such a database is needed to analyze past experiences and replicate successes.

Recommendation 9: Public information and protection—A centralized system is needed to provide information to the public on air quality impacts from fires. Also needed are general criteria for how land managers, air regulators, and public health officials should respond to adverse smoke impacts and emergency notifications of the public to health hazards associated with smoke from fire.

The authors of this plan hoped that these recommendations would be used in future joint agency efforts to advance the fire sciences, minimize duplication of effort, and share information among agencies and the public.

The technically advanced smoke estimation tools, or TASET, project (Fox and Riebau 2000) was funded by the Joint Fire Sciences Program (JFSP) to develop a structured analysis of smoke management and recommend specific developments for advancing the state of science. The report confirmed and refined the recommendations of Sandberg and others (1999) above, and developed 10 recommendations for research activities:

- Fire community participation in regional air quality modeling consortia.
- Conduct a national smoke and visibility conference and reference guide.
- Develop a national smoke emissions data structure or database system.
- Apply remote sensing for fuels and fire area emissions inventories.
- Develop a fire gaming system to quantify emissions and impacts from alternative fire management practices.
- Improve the CalMet/CalPuff smoke management model.
- Upgrade a nationalized screening model/simple approach smoke estimation model (SASEM).
- Provide onsite fire emissions verification.
- Utilize back-trajectory modeling and filter analysis for fire smoke contributions for nonattainment areas.
- Develop a method to identify the specific sources of organic carbon fine particulate material.

Research priorities established in the *Effects of Fire Air* (Sandberg and others 1979) are unfortunately still valid today, although some progress has been made in every category. We list these here, slightly reworded from the original for brevity and to conform to modern nomenclature:

1. Provide quantitative smoke management systems.
   a. Develop information systems necessary to support smoke management decisions.
   b. Provide a smoke management reporting system for emission rates based on the prediction of fuel consumption, fire behavior, heat release rates, and source control measures.
   c. Provide the data network and modeling scheme to calculate the change in pollution concentrations and character between the source and potential receptors.
   d. Adapt plume rise models necessary to predict the vertical distribution of emissions from fires.
2. Characterize the chemistry and physics of emissions.
   a. Relate emissions and heat release rates to fuelbed characteristics and fire behavior.

64

USDA Forest Service Gen. Tech. Rep. RMRS-GTR-42-vol. 5. 2002

b. Advance our knowledge of hazardous and reactive compounds in smoke.

c. Develop field methods to monitor emission rates and smoke chemistry from operational fires.

d. Investigate the potential for secondary reactions of emissions downstream from their source.

3. Model atmospheric transport, diffusion, transformation, and removal mechanisms.

a. Continue development of winds and dispersion models for boundary layer flow and mesoscale transport of smoke over mountainous terrain.

b. Investigate the mechanisms of removal; for example, canopy interactions, fallout, and local deposition.

c. Interact with the wider scientific community to establish the effect of reactive pollutants on the biosphere.

d. Evaluate the potential contribution of wildland fires to climate change.

4. Identify receptor responses to wildland smoke.

a. Identify and quantify the visibility needs of wildland users, and recommend standards for particulate and sulfate pollution from all sources affecting Class I visibility areas.

b. Evaluate the potential impact of wildland smoke on human health.

c. Investigate the role of wildland ecosystems as a sink and receptor for atmospheric contaminants.

5. Investigate tradeoffs made in the substitution of alternatives to fire use.

a. Develop simulation models to evaluate interactions of land use policy with air resource management. Incorporate air resource management and fuels management needs into the land use planning process.

b. Evaluate the effect on wildland fire occurrence and air pollution from changes in the amount of prescribed fire activity.

c. Describe the resource and economic tradeoff of wildland fire occurrence resulting from a change in prescribed fire activity.

d. Investigate the effect of changes in fire use on nutrient cycling, successional response, and ecosystem stability.

Emerging Research Needs _____

Several new responsibilities create the need for additional information systems that require new research and development, including:

- Planning rules that require the consideration of cumulative pollution and visibility impacts of fuel management programs.
- Wildland fire situation analysis requirements that smoke impacts from wildland be anticipated and communicated to the public.
- Increased requirements for emission reduction.
- Policies that require hourly and daily tracking of emissions and the management of smoke from all fires.
- Increased management of wildland fires for resource benefits.
- Increased use of long-duration landscape-scale fires.
- Regulatory concern over secondary pollutants, especially ozone formation and the reentrainment of mercury.
- Questions about the role of fire and global biomass emission on atmospheric carbon and global warming.
- Increased attention to firefighter health effects from exposure to smoke.

Each of these factors requires information systems for planning, operations, and monitoring the effects of fire on air. Using the framework illustrated in figures 1-1 and 1-2 (in chapter 1) and the background of previous chapters, some emerging research needs are outlined below.

Emissions Source Strength and Emissions Inventory

Level of burn activity: Accurately predict, determine, and record the area burned and time of burning for all types of prescribed and wildland fire—Area burned is still the parameter that imparts the greatest error into predictions of source strength and emission inventory. Needed are: a balanced program of new planning models that project area burned and fire residence times; remote-sensing technologies that track fire sizes at hourly intervals; ground based sampling, reporting, and communication systems; and analysis tools. Planning models include those that project fire use and predict wildland fire activity from 1 to 50 years in the future must be included, as well as accurate predictions made a day in advance.

Biomass: Accurately predict, determine, and record the mass, combustion stage, and residence time of fuels burned in all types of fires—Inadequate representation of fuelbed characteristics and the ability to infer fuelbed characteristics and flammability conditions from remote sensing or ecosystem physiognomy is the second greatest remaining source of error. Models of the combustion process,

USDA Forest Service Gen. Tech. Rep. RMRS-GTR-42-vol. 5. 2002

65

while improving, are still inadequate to predict or characterize emission rates and durations. New classification systems, inference models, inventory and sensing processes, and process models are needed.

Heat release and emissions: Predict and measure physical and chemical characteristics of emissions from all types of fires—Among the greatest advances since about 1980 has been the nearly complete characterization of primary and criteria pollutants from a wide range of fire environments. New models also greatly improve the prediction and characterization of emissions source strength. Emission factors for criteria pollutants are adequate. There is substantial remaining uncertainly in the measurement and prediction of precursors to ozone and other secondary chemical formations, secondary entrainment of mercury, production and stimulation of nitrogen compounds, air toxics, and greenhouse gases. Continuing research on these trace constituents are needed. In addition, we lack models that characterize the complex spatial and temporal distribution of heat release from fires.

Emissions inventory methods: Integrate measurements and reporting from remote sensing, airborne platforms, simulation models, and surface observations into a fine-scale spatial and temporal emission inventory—Emission inventories are a fundamental tool that air resource managers use to calculate the relative importance of air pollution sources and to design control strategies. Hourly, point-specific emission estimates as well as daily, monthly, and yearly summaries are necessary to compare fire with other sources or as inputs to dispersion models. Fire managers currently lack a system of observations and reporting mechanisms required for planning, tracking, and monitoring emissions.

Ambient Air Quality Impacts

Background air quality: Improve the accessibility of girded detail about background air quality and meteorological conditions—Fire emissions are inserted into an already complex atmosphere, and current ability to predict pollutant interactions, transformations, and combined effects are limited by the availability of hourly fine-scale atmospheric profiling.

Plume rise and transport: Improve the prediction, detection, and tracking of plumes from all types and stages of fires—Fire plumes are complex; often splitting into lofted and unlofted portions; plumes that split in two directions at different altitudes, and plumes that change rapidly over time. Plumes are transported long distances, often over complex terrain, and the accuracy and availability of models to predict transport are inadequate. Methods to track plume trajectories and measure pollutant concentrations in near real time using remote sensing are emerging but not yet available.

Dispersion, dilution, and pollutant transformation: Improve the ability on all scales to predict, model, and detect changes in the properties and concentration of pollutants over time and space—Data and models are needed to initiate and predict local, regional, national, and global air quality impacts from individual fires to the cumulative effects of tens of thousands of fires.

Atmospheric carbon balance and climatic change: Develop consistent technologies to assess the contribution of fires to greenhouse gases in the atmosphere and the effect of fire and ecosystem management practices—For a source of greenhouse gas emissions as large as wildland and prescribed fires, there is a regrettable lack of consensus on the magnitude or even the methods for assessment and accountability. This emerging issue requires much of the same research on source characteristics and air quality as do the health, safety, and visibility issues, but also requires integration with the global science and policy communities.

Effects on Receptors

Visibility and other welfare effects: Predict, measure, and interpret the impact of natural and anthropogenic fire sources on visibility, economic, and other welfare effects—The impact of smoke exposure from fires on human health standards is minor relative to the nuisance it creates and the impacts on visibility. New science is required to monitor and predict effects on visibility, and to apportion visibility impacts to specific sources and classes of sources.

Health and safety risk assessment: Develop knowledge and systems to assess the risk of individual and collective fires to personal and community health and safety—This broad topic has received limited attention in recent years, mostly in the prediction of visibility impacts on highway safety and in the assessment of individual firefighter exposure to hazardous air pollutants. But all aspects of risk management, including hazard identification, exposure assessment, dose-response, risk assessment, and mitigation measures are lacking.

66

USDA Forest Service Gen. Tech. Rep. RMRS-GTR-42-vol. 5. 2002

Conclusion

Knowledge and information requirements for managing fire effects on air quality continue to increase. Policy advancements require the understanding, modeling, prediction, monitoring, and tracking of fires and their effect on air at greater detail and in greater volume than ever before. Research and development has progressed logically over the past 25 years due to strategic planning and prioritization that has included the needs of the managers of ecosystems and of air quality. Analytical and information transfer capacity has increased dramatically in the past decade, so information is more readily accessible to those who need it. Thanks largely to the National Fire Plan, the Joint Fire Science Program, the Western Regional Air Partnership, and EPA's implementation of the Regional Haze Rule, there is currently more active research and development the effects of fire on air than ever before.

USDA Forest Service Gen. Tech. Rep. RMRS-GTR-42-vol. 5. 2002

67

Notes

_____

_____

_____

_____

_____

_____

References

40 CFR Part 51. Vol. 64 No. 126. Regional Haze Regulations—Final Rule. July 1, 1999.

Achtemeier, G.L. 1994. A computer wind model for predicting smoke movement. Southern Journal of Applied Forestry. 18: 60–64.

Achtemeier, Gary L. 2000. PB-Piedmont: A numerical model for predicting the movement of biological material near the ground at night. In: Proceedings of the 24th conference on agricultural and forest meteorology. Boston, MA: American Meteorology Society: 178–179.

Achtemeier, Gary L. 2001. Simulating nocturnal smoke movement. Fire Management Today. 61: 28–33.

Albini, F.A.; Brown, J.K.; Reinhardt, E.D.; Ottmar, R.D. 1995. Calibration of a large fuel burnout model. International Journal of Wildland Fire. 5(3): 173–192.

Albini, F.A.; Reinhardt, E.D. 1995. Modeling ignition and burning rate of large woody natural fuels. International Journal of Wildland Fire. 5(2): 81–91.

Albini, F.A.; Reinhardt, E.D. 1997. Improved calibration of a large fuel burnout model. International Journal of Wildland Fire. 7(1): 21–28.

Anderson, Hal E. 1969. Heat transfer and fire spread. Res. Pap. INT-69. Ogden, UT: U.S. Department of Agriculture, Forest Service, Intermountain Forest and Range Experiment Station. 20 p.

Andreae, M.O.; Browell, E.V.; Garstang, M.; Gregory, G.L.; Harriss, R.C.; Hill, G.F.; Jacob, D.J.; Pereira, M.C.; Sachse, G.W.; Setzer, A.W.; Silva Dias, P.L.; Talbot, R.W.; Torres, A.L.; Wofsy, S.C. 1988. Biomass-burning emissions and associated haze layers over Amazonia. Journal of Geophysical Research. 93: 1509–1527.

Andreae, M.O.; Anderson, B.E.; Blake, D.R.; Bradshaw, J.D.; Collins, J.E.; Gregory, G.L.; Sachse, G.W.; Shipman, M.C. 1994. Influence of plumes from biomass burning on atmospheric chemistry over the equatorial and tropical South Atlantic during CITE 3. Journal of Geophysical Research. 99(D6): 12,793–12,808.

Andreae, M.O.; Merlet, P. 2001. Emission of trace gases and aerosols from biomass burning. Global Biogeochemical Cycles. 15(4): 955–966.

Andrews, Patricia L.; Bevins, Collin D. 1999. BEHAVE fire modeling system—redesign and expansion. Fire Management Notes. 59(2): 16–19.

Baedecker, P.A.; Edney, E.O.; Morgan, P.J.; Simpson, T.C.; Williams, R.S. 1991. Effects of acidic deposition on materials. In: Irving, P.M., ed. Acidic deposition: state of science and technology, volume III: terrestrial, materials, health and visibility effects. Washington, DC: The U.S. National Acidic Precipitation Assessment Program (NAPAP). Atmospheric Environment. 26: 147–158.

Battye, R.; Bauer, B.; MacDonald, G. 1999. Features of prescribed fire and smoke management rules for western and southern States. Contract 68-D-98-026. Chapel Hill, NC: EC/R Incorporated. Prepared for U.S. Environmental Protection Agency.

Battye, William; Battye, Rebecca. 2002. Development of emissions inventory methods for wildland fire. Final report. Contract 68-D-98-046. Research Triangle Park, NC: U.S. Environmental Protection Agency.

Beason, Donald G.; Johnson, James S.; Foote, Kenneth L.; Weaver, William A. 1996. Summary report. California Department of Forestry and Fire Protection evaluation of full-face air-purifying respirators for wildland fire fighting use. California Department of Forestry and Fire Protection Contract WN-02-19-05-0. Livermore, CA: Lawrence Livermore National Laboratory. February.

Beloin, N.J.; Haynie, F.H. 1975. Soiling of building materials. Journal of the Air Pollution Control Association. 25: 399–403.

Betchley, C.; Koenig, J.Q.; van Belle, G. [and others]. 1995. Pulmonary function and respiratory symptoms in forest firefighters. Unpublished report. On file with: University of Washington, Departments of Environmental Health and Epidemiology, Seattle, WA.

Booze, Thomas F.; Reinhardt, Timothy E. [In press]. A screening-level assessment of the health risks of chronic smoke exposure for wildland firefighters. American Industrial Hygiene Association Journal.

Bradley, Michael M.; Schomer, Christina L.; Sumikawa, Denise A.; Walker, Hoyt; Younker, Leland W.; Bossert, James E.; Hanson, Howard P.; Linn, Rodman R.; Reisner, Jon M. 2000. The national wildfire prediction program: a key piece of the wildfire solution. In: Neuenschwander, Leon, F.; Ryan, Kevin C., tech. eds. Proceedings from the joint fire science conference and workshop: crossing the millennium: integrating spatial technologies and ecological principles for a new age in fire management. University of Idaho: 64–76.

Brenner, J.; Suddaby, R.M.; Carr, R.J.; Lee, B.S.; Brackett, D.P.; Arvanitus, L.G. 2001. GIS-based fire management in Florida. Journal of Forestry. 95(6): 140–147.

Breyfogle, Steve; Ferguson, Sue A. 1996. User assessment of smoke-disperion models for wildland biomass burning. Gen. Tech. Rep. PNW-GTR-379. Portland, OR: U.S. Department of Agriculture, Forest Service, Pacific Northwest Research Station. 30 p.

Briggs, G.A. 1969. Plume rise. Prepared for: Nuclear Safety Information Center, Oak Ridge National Laboratory. Oak Ridge, TN: U.S. Atomic Energy Commission, Division of Technical Information. Springfield, VA: Clearinghouse for Federal Scientific and Technical Information, National Bureau of Standards, U.S. Dept. of Commerce.

Brookshire, D.S.; Ives, B.C.; Schulze, W.D. 1976. The valuation of aesthetic preferences. Journal of Environmental Economics and Management. 3: 325–346.

Brown, Arthur A.; Davis, Kenneth P. 1959. Forest fire: control and use. New York: McGraw-Hill Book Company. 686 p.

Brown, J.K.; Reinhardt, E.D.; Fischer, W.C. 1991. Predicting duff and woody fuel consumption in northern Idaho prescribed fires. Forest Science. 37(6): 1550–1566.

Bryan, Dana C., ed. 1997. Conference proceedings: Environmental regulation & prescribed fire: legal and social challenges. Tallahassee, FL: Florida State University, Center for Professional Development. 246 p.

Byun, D.W. 1999a. Dynamically consistent formulations in meteorological and air quality models for multiscale atmospheric studies. Part I: Governing equations in a generalized coordinate system. Journal of Atmospheric Sciences. 56: 3789–3807.

Byun, D.W. 1999b. Dynamically consistent formulations in meteorological and air quality models for multiscale atmospheric studies. Part II: Mass conservation issues. Journal of Atmospheric Sciences. 56: 3808–3820.

Byun, D.W.; Ching, J.K.S., eds. 1999. Science algorithms of the EPA Models-3 community multiscale air quality (CMAQ) modeling system. EPA/600/R-99/030. U.S. Environmental Protection Agency, Office of Research and Development.

Byun, D.W.; Pleim, J.E.; Tang, R.T.; Bourgeois, A. 1999. Chapter 12: meteorology-chemistry interface processor (MCIP) for MODELS-3 community multiscale air quality (CMAQ) modeling system. In: Byun, D.W.; Ching, J.K.S., eds. Science algorithms of the EPA Models-3 community multiscale air quality (CMAQ) modeling system. EPA/600/R-99/030. U.S. Environmental Protection Agency, Office of Research and Development.

Cassinelli, M.E.; O'Connor, P.F. eds. 1994. NIOSH manual of analytical methods, 4th ed. Washington, DC: U.S. Government Printing Office.

Chameides, W.L.; Cowling, E.B. 1995. The state of the Southern Oxidants Study (SOS): policy-relevant findings in ozone pollution research, 1988–1994. Raleigh, NC: North Carolina State University, College of Forestry. 94 p.

Chang, J.S.; Chang, K.H.; Jin, S. 1993. Two-way and one-way nested SARMAP air quality model. In: International conference on regional photochemical measurement and modeling studies. Pittsburgh, PA: Air & Waste Management Association.

USDA Forest Service Gen. Tech. Rep. RMRS-GTR-42-vol. 5. 2002

69

Chatfield, R.B.; Delaney, A.C. 1990. Convection links biomass burning to increased tropical ozone: however, models will tend to over predict O_3. Journal of Geophysical Research. 95: 18,473–18,488.

Chatfield, R.B.; Vastano, J.A.; Singh, H.B.; Sachase, G.W. 1996. A general model of how fire emissions and chemistry produce African/oceanic plumes (O_3, CO, PAN, smoke) in TRACE A. Journal of Geophysical Research. 101(D19): 24,279–24,306.

Chestnut, L.G.; Rowe, R.D. 1990. Preservation values for visibility in the National Parks. Washington, DC: U.S. Environmental Protection Agency.

Chi, C.T.; Horn, D.A.; Reznik, R.B.; [and others]. 1979. Source assessment: prescribed burning, state of the art. Final report. EPA-600/2-79-019h. Research Triangle Park, NC: U.S. Environmental Protection Agency, Office of Research and Development. 106 p.

Chow, J.C. 2000. Comparison of IMPROVE and NIOSH carbon measurements. Presented at the PM2000: Particulate matter and health conference. Pittsburg, PA: Air & Waste Management Association.

Clark, T.L.; Jenkins, M.A.; Coen, J.; Packham, D. 1996. A coupled atmosphere-fire model: convective feedback on fire-line dynamics. Journal of Applied Meteorology. 35(6): 875–901.

Clean Air Scientific Advisory Committee. 1995. Clean Air Scientific Advisory Committee (CASAC) Comments on the April 1995 draft air quality criteria for particulate matter. EPA-SAB-CASAC-LTR-95-005. U.S. Environmental Protection Agency, Science Advisory Committee, Clean Air Scientific Advisory Committee, Particulate Matter Criteria Document Review Panel. 8 p.

Core, J.E. 1989. Air quality and forestry burning: public policy issues. In: Hanley, D.P.; Kammenga, J.J.; Oliver, C.D., eds. The burning decision: regional perspectives on slash. Seattle, WA: University of Washington, College of Forest Resources: 237–245.

Core, J.E. 1996. Wildfire smoke emergency action plan implementation guideline: draft of 11/15/95. In: WESTAR Council wildfire and prescribed fire workshop. Portland, OR: WESTAR Council.

Core, J.E. 1998. Survey of smoke management programs in the western United States. Prepared for the USDI Bureau of Land Management, National Applied Resources Center. May.

Cowling, J.E.; Roberts, M.E. 1954. Paints, varnishes, enamels, and lacquers. In: Greathouse, G.A.; Wessel, C.J., eds. Deterioration of materials: causes and preventive techniques. New York: Reinhold Publishing Corporation: 596–645.

Crutzen, P.J.; Andreae, M.O. 1990. Biomass burning in the tropic: impacts on atmospheric chemistry and biogeochemical cycles. Science. 250(4988): 1669–1678.

Crutzen, P.J.; Carmichael, G.J. 1993. Modeling the influence of fires on atmospheric chemistry. In: Crutzen, P.J.; Goldammer, J.G., eds. Fire in the environment: the ecological, atmospheric, and climatic importance of vegetation fires. New York: Wiley: 89–105.

Davies, Mary Ann. 2002. DataRAM4 particulate monitor: Forest Service user's guide. Tech. Rep. 0225-2810-MTDC. Missoula, MT: U.S. Department of Agriculture, Forest Service, Technology and Development Program. 26 p.

DeBano, Leonard F.; Neary, Daniel G.; Ffolliott, Peter F. 1998. Fire's effects on ecosystems. New York: John Wiley and Sons, Inc. 333 p.

de Nevers, N. 2000. Air pollution control engineering. Boston, MA: McGraw Hill. 586 p.

Desalmand, F.; Serpolay, R.; Podzimek, J. 1985. Some specific features of the aerosol particle concentrations during the dry season and during a bush fire event in West Africa. Atmospheric Environment. 19(9): 1535–1543.

Dickson, R.J.; Oliver, W.R.; Dickson, E.L. 1994. Development of an emissions inventory for assessing visual air quality in the western United States. Prepared for the Western Governors' Association and the Electric Power Research Institute. Radian Corporation 674-050-04-01. July.

Dockery, D.W.; Pope, C.A., III; Xu, X.; Spengler, J.D.; Ware, J.H.; Fay, M.E.; Ferris, B.G., Jr.; Speizer, F.E. 1993. An association between air pollution and mortality in six U.S. cities. New England Journal of Medicine. 329: 1753–1759.

Draxler, R.R.; Hess, G.D. 1998. An overview of the HYSPLIT-4 modelling system for trajectories, dispersion and deposition. Australian Meteorological Magazine. 47(4): 295–308.

Dull, K.; Acheson, A.; Thomas, D.; Chapell, L.; Volkland, S. 1998. Forecasting smoke dispersion and concentration in Idaho and Montana. Unpublished report. On file at: U.S. Department of Agriculture, Forest Service, Pacific Northwest Region, Portland, OR. 13 p.

Eatough, D.J.; Pang, Y. 1999. Determination of PM2.5 sulfate and nitrate with a PC-BOSS designed for routine sampling for semi-volatile particulate matter. Journal of the Air & Waste Management Association. 49: 69–75.

Eshee, W.D. 1995. Legal implications of using prescribed fire. In: Bryan, D.C., ed. Proceedings: environmental regulation and prescribed fire conference: legal and social challenges. Tallahassee, FL: Division of Forestry, Florida Department of Agriculture and Consumer Services: 126–130.

Ferguson, S.A.; Peterson, J.; Acheson, A. 2001. Automated, real-time predictions of cumulative smoke impacts from prescribed forest and agricultural fires. In: Fourth symposium on fire and forest meteorology. Boston, MA: American Meteorological Society: 168–175.

Ferguson, Sue A.; Sandberg, David V.; Ottmar, Roger. 2000. Modelling the effect of landuse changes on global biomass emissions. In: Innes, John L.; Beniston, Martin; Verstraete, Michel M., eds. Biomass burning and its relationships with the climate system. Dordrecht, The Netherlands: Kluwer Academic Publishers: 33–50.

Finlayson-Pitts, Barbara J.; Pitts, James N., Jr. 1986. Atmospheric chemistry: fundamentals and experimental techniques. New York: John Wiley and Sons, Inc. 1098 p.

Finney, M. 1998. FARSITE: Fire area simulator—model development and evaluation. Res. Pap. RMRS-RP-4. Ogden, UT: U.S. Department of Agriculture, Forest Service, Rocky Mountain Research Station. 47 p.

Finney, Mark. 2000. [Personal communication]. Research forester. Rocky Mountain Research Station, Fire Sciences Laboratory, Missoula, MT.

Fire Emissions Joint Forum. 2002. WRAP policy: enhanced smoke management programs for visibility. http://www.wrapair.org/forums/FEJF/esmptt/policy/WRAP_ESMP_Policy_090202.pdf. [September 2002].

Fishman, J.; Fakhruzzaman, K.; Cros, B.; Ngana, D. 1991. Identification of widespread pollution in the Southern Hemisphere deduced from satellite analyses. Science. 252: 1693–1696.

Fox, Douglas G.; Riebau, Allen R. 2000. Technically advanced smoke estimation tools (TASET). Final report. Fort Collins, CO: Colorado State University, Cooperative Institute for Research in the Atmosphere. 99 p.

Goode, J.G.; Yokelson, R.J.; Susott, R.A.; Ward, D.E. 1999. Trace gas emissions from laboratory biomass fires measured by open-path Fourier transform infrared spectroscopy: fires in grass and surface fuels. Journal of Geophysical Research. 104: 21,237–21,245.

Goode, J.G.; Yokelson, R.J.; Susott, R.A.; Babbitt, R.E.; Ward, D.E.; Davies, M.A.; Hao, W.M. 2000. Measurements of excess O_3, CO_2, CO, CH_4, C_2H_4, C2H2, HCN, NO, NH_3, HCOOH, CH_3COOH, HCHO, and CH_3OH in 1997 Alaskan biomass burning plumes by airborne Fourier transform infrared spectroscopy (AFTIR). Journal of Geophysical Research. 105: 22,147.

Grand Canyon Visibility Transport Commission. 1996a. Alternative assessment committee report. Denver, CO: Western Governors' Association.

Grand Canyon Visibility Transport Commission. 1996b. Recommendations for improving western vistas. Denver, CO: Western Governors' Association.

Grand Canyon Visibility Transport Commission. 1996c. Report of the Grand Canyon Visibility Transport Commission to the United States Environmental Protection Agency (1996). Denver, CO: Western Governors' Association. 85 p.

Grell, G.A.; Dudhia, J.; Stauffer, D.R. 1995. A description of the fifth-generation Penn State/NCAR mesoscale model (MM5). NCAR Technical Note, NCAR/TN-398+ STR. Boulder, CO: National Center for Atmospheric Research.

Haines, D.A.; Smith, M.C. 1987. Three types of horizontal vortices observed in wildland mass and crown fires. Journal of Climate and Applied Meteorology. 26: 1624–1637.

Haines, Donald A.; Updike, Gerald H. 1971. Fire whirlwind formation over flat terrain. Res. Pap. NC-71. St. Paul, MN: U.S.

Department of Agriculture, Forest Service, North Central Forest Experiment Station. 12 p.

Hao, Wei Min; Lui, Mei-Huey. 1994. Spatial and temporal distribution of tropical biomass burning. Global Biogeochemical Cycles. 8(4): 495–503.

Hardy, C.C.; Conard, S.G.; Regelbrugge, J.C.; Teesdale, D.T. 1996. Smoke emissions from prescribed burning of southern California chaparral. Res. Pap. PNW-RP-486. Portland, OR: U.S. Department of Agriculture, Forest Service, Pacific Northwest Research Station. 37 p.

Hardy, C.C.; Ferguson, S.A.; Speers-Hayes, P.; Doughty, C.B.; Teasdale, D.R. 1993. Assessment of PUFF: a dispersion model for smoke management. Final report. Submitted to: U.S. Department of Agriculture, Forest Service, Pacific Northwest Region. 32 p.

Hardy, C.C.; Menakis, J.P.; Long, D.G.; Garner, J.L. 1998. FMI/ WESTAR emissions inventory and spatial data for the Western United States. Final report. EPA agreement number DW12957250-01-0. U.S. Department of Agriculture, Forest Service, Rocky Mountain Research Station, Fire Effects Research Work Unit, Missoula, MT.

Hardy, C.C.; Ward, D.E.; Enfield, W. 1992. PM2.5 emissions from a major wildfire using a GIS: rectification of airborne measurements. In: Proceedings, 29th annual meeting of the Pacific Northwest International Section, Air & Waste Management Association. Pittsburgh, PA: Air & Waste Management Association.

Hardy, Colin C.; Ottmar, Roger D.; Peterson, J.L.; Core, John E.; Seamon, Paula, comps., eds. 2001. Smoke management guide for prescribed and wildland fire: 2001 edition. PMS 420-2. Boise, ID: National Wildfire Coordinating Group. 226 p.

Harms, Mary F.; Lavdas, Leonidas G. 1997. Draft user's guide to VSMOKE-GIS for workstations. Research Paper SRS-6. Asheville, NC: U.S. Department of Agriculture, Forest Service, Southern Research Station. 41 p.

Harrison, H. 1995. A user's guide to NFSPUFF: a dispersion model for smoke management in complex terrain. WYNDSoft Inc. 42 p.

Haynie, F.H.; Lemmons, T.J. 1990. Particulate matters oiling of exterior households paints. Journal of the Air Pollution Control Association. 34: 941–944.

Haynie, F.H.; Spence J.W.; Lipfert, F.W. Cramer, S.D.; McDonald, L.G. 1990. Evaluation of an atmospheric damage function for galvanized steel. In: Baboian, R.; Dean, S.W., eds. Corrosion testing and evaluation: silver anniversary volume. ASTM Special Tech. Publ. 1000. Philadelphia, PA: American Society for Testing and Materials: 225–240.

Hobbs, P.V.; Radke, L.F. 1969. Cloud condensation nuclei from a simulated forest fire. Science. 163: 279–280.

Hobbs, P.V.; Reid, J.S.; Herring, J.A.; Nance, J.D.; Weiss, R.E.; Ross, J.L.; Hegg, D.A.; Ottmar, R.D.; Liousse, C. 1996. Particle and trace-gas measurements in the smoke from prescribed burns of forest products in the Pacific Northwest. In: Levine, J.S., ed. Biomass burning and global change: volume I remote sensing, modeling and inventory development, and biomass burning in Africa. Cambridge, MA: The MIT Press: 697–715.

Hopke, P.K. 1985. Receptor modeling in environmental chemistry. New York: Wiley.

Hummel, J.; Rafsnider, J. 1995. User's Guide, TSARS plus smoke production and dispersion model. Unpublished. National Biological Service and the Interior Fire Coordination Committee. 107 p.

Idaho Department of Environmental Quality. [n.d.] Supporting technical document for PM10 excursions in Salmon, Idaho, during the summer of 2000. Boise, ID: Idaho Department of Environmental Quality. http://www2.state.id.us/deq/air/smoke/NEAP/ NEAP_SupportingDoc.pdf. (August 8, 2002).

Iowa Supreme Court. 1998. Bormann and others versus Board of Supervisors in and for Kossuth County, Iowa. September 23, 1998. No. 192/96-2276. http://www.judicial.state.ia.us/supreme/ opinions/19980923/96-2276.asp. (August 8, 2002).

Jacob, D.J.; Heikes, B.G.; Fan, S.-M.; Logan, J.A.; Mauzerall, D.L.; Bradshaw, J.D.; Singh, H.B.; Gregory, G.L.; Talbot, R.W.; Blake, D.R.; Sachse, G.W. 1996. Origin of ozone and NOx in the tropical troposphere: a photochemical analysis of aircraft observations over the South Atlantic Basin. Journal of Geophysical Research. 101(24): 24,235–24,250.

Jacob, D.J.; Wofsy, S.C.; Bakwin, P.S.; Fan, S.-M.; Harriss, R.C.; Talbot, R.W.; Bradshaw, J.D.; Sandholm, S.T.; Singh, H.B.; Browell, E.V.; Gregory, G.L.; Sachse, G.W.; Shipham, M.C.; Blake, D.R.; Fitzjarrald, D.R. 1992. Summertime photochemistry of the troposphere at high northern latitudes. Journal of Geophysical Research. 97: 16,421–16,431.

Jennings, S.G. 1998. Wet processes affecting atmospheric aerosols. In: Harrison, R.M.; Grieken, R.V., eds. Atmospheric particles: 475–508.

Kasischke, E.S.; Stocks, B.J., eds. 2000. Fire, climate change, and carbon cycling in the boreal forest. Ecological Studies, Vol. 138. New York: Springer. 461 p.

Khalil, M.A.K.; Edgerton, S.A.; Rasmussen, R.A. 1983. A gaseous tracer model for air pollution from residential wood burning. Environmental Science and Technology. 22: 53–61.

Koppmann, R.; Khedim, A.; Rudolph, J.; Poppe, D.; Andreae, M.O.; Helas, G.; Welling, M.; Zenker, T. 1997. Emissions of organic trace gases from savanna fires in southern Africa during the 1992 Southern Africa Fire Atmosphere Research Initiative and their impact on the formation of tropospheric ozone. Journal of Geophysical Research. 102: 18,879–18,888.

Kumar, N.; Russell, A.G. 1996. Development of a computationally efficient, reactive sub-grid scale plume model and the impact in the northeastern United States using increasing levels of chemical detail. Journal of Geophysical Research. 101: 16,737–16,744.

Lavdas, L.G. 1982. A day/night box model for prescribed burning impact in Willamette Valley, Oregon. Journal of Air Pollution Control Agency. 32: 72–76.

Lavdas, L.G. 1996. Program VSMOKE—user manual. Gen. Tech. Rep. SRS-6. Asheville, NC: U.S. Department of Agriculture, Forest Service, Southern Research Station. 147 p.

Laverty, Lyle; Williams, Jerry. 2000. Protecting people and sustaining resources in fire-adapted ecosystems: a cohesive strategy. Washington, DC: U.S. Department of Agriculture, Forest Service. 85 p.

Lavoué, David; Liousse, Catherine; Cachier, Hélène. 2000. Modeling of carbonaceous particles emitted by boreal and temperate wildfires at northern latitudes. Journal of Geophysical Research. 105(D22): 26,871–26,890.

Lee, M.; Heikes, B.G.; Jacob, D.J. 1998. Enhancements of hydroperoxides and formaldehyde in biomass burning impacted air and their effect on atmospheric oxidant cycles. Journal of Geophysical Research. 13,201–13,212.

Leenhouts, Bill. 1998. Assessment of biomass burning in the conterminous United States. Conservation Ecology [online]. 2(1). http://www.consecol.org/vol2/iss1/art1 (August 8, 2002).

Lelieveld, J.; Crutzen, P.J.; Jacob, D.J.; Thompson, A.M. 1997. Modeling of biomass burning influences on tropospheric ozone. In: Wilgen, B.W., ed. Fire in southern African savannas: ecological and atmospheric perspectives. Johannesburg, South Africa: Witsatersrand University Press: 217–238.

Levin, J.S., ed. 1996. Global biomass burning: Atmospheric, climatic, and biospheric implications. Cambridge, MA: MIT Press. 569 p.

Levine, Joel S.1994. Biomass burning and the production of greenhouse gases. In: Zepp, Richard G. Climate-biosphere interactions: biogenic emissions and environmental effects of climate change. New York: John Wiley and Sons, Inc.: 139–160.

Levine, Joel S.; Cofer, Wesley R., III. 2000. Boreal forest fire emissions and the chemistry of the atmosphere. In: Kasischke, Eric S.; Stocks, Brian J., eds. Fire, climate change, and carbon cycling in the boreal forest. Ecological Studies Vol. 138. New York: Springer-Verlag: 31-48.

Levinson, D. 2001. The Montana/Idaho airshed group operating guide. Missoula, MT: Montana/Idaho Airshed Group, Smoke Monitoring Unit. 48 p.

Ligocki, M.P.; Salmon, L.G.; Fall, T.; Jones, M.C.; Nazaroff, W.W.; Cass, G.R. 1993. Characteristics of airborne particles inside southern California museums. Atmospheric Environment. Part A. 27: 697–711.

Lipfert, F.W.; Daum, M.L. 1992. The distribution of common construction materials at risk to acid deposition in the United States. Atmospheric Environment Part B. 26: 217–226.

USDA Forest Service Gen. Tech. Rep. RMRS-GTR-42-vol. 5. 2002

71

Little, Beth. [n.d.]. CalPFIRS [California prescribed fire incident reporting system] program manual. Redding, CA: U.S. Department of Agriculture, Forest Service, Northern California Service Center. 43 p.

Lobert, J.M.; Scharffe, D.H.; Hao, W.M.; Kuhlbusch, T.A.; Seuwen, R.; Warneck, P.; Crutzen, P.J. 1991. Experimental evaluation of biomass burning emissions: Nitrogen carbon containing compounds. In: Levin, J.S., ed. Global biomass burning: Atmospheric, climatic, and biospheric implications. Cambridge, MA: MIT Press. 569 p.

Maler, K.G.; Wyzga, R.E. 1976. Economic measurements of environmental damage: a technical handbook. Paris: Organization for Economic Cooperation and Development.

MacFarland, K.K.; Malm, W.; Molenar, J. 1983. An examination of methodologies for assessing the value of visibility. In: Rowe, R.D.; Chestnut, L.G., eds. Managing air quality and scenic resources at National Parks and wilderness areas. Boulder, CO: Westview Press: 151–172.

Malm, W.C. 2000. Introduction to visibility. #CA-2350-97. Ft. Collins, CO: Colorado State University, Cooperative Institute for Research in the Atmosphere: T097-04, T098-06.

Materna, B.L.; Jones, J.R.; Sutton, P.M. [and others]. 1992. Occupational exposures in California wildland fire fighting. American Industrial Hygiene Association Journal. 53(1): 69–76.

Martins, J.V.; Artaxo, P.; Hobbs, P.V.; Liousse, C.; Cachier, H.; Kaufman, Y.; Plana-Fattori, A. 1996. Particle size distributions, elemental compositions, carbon measurements, and optical properties of smoke from biomass burning in the Pacific Northwest of the United States. In: Levin, J.S., ed. Biomass burning and global change, volume 2: Biomass burning in South America, southeast Asia, and temperate and boreal ecosystems, and the oil fires of Kuwait. Cambridge, MA: The MIT Press: 716–732.

Mathur, R.; Peters, L.K.; Saylor, R.D. 1992. Sub-grid presentation of emission source clusters in regional air quality monitoring. Atmospheric Environment. 26A: 3219–3238.

Mauzerall, D.L.; Jacob, D.J.; Fan, S.M.; Bradshaw, J.D.; Gregory, G.L.; Sachse, G.W.; Blake, D.R. 1996. Origin of tropospheric ozone at remote high northern latitudes in summer. Journal of Geophysical Research. 101(D2): 4175–4188.

Mauzerall, D.L.; Logan, J.A.; Jacob, D.J.; Anderson, B.E.; Blake, D.R.; Bradshaw, J.D.; Heikes, B.; Sachse, G.W.; Singh, H.; Talbot, R. 1998. Photochemistry in biomass burning plumes and implications for tropospheric ozone over the tropical South Atlantic. Journal of Geophysical Research. 103: 8401–8423.

McClelland, G.; Schulze, W.; Waldman, D.; Irwin, J.; Schenk, D.; Stewart, T.; Deck, L.; Thayer, M. 1991. Valuing eastern visibility; a field test of the contingent valuation method. Cooperative agreement CR-815183-01-3. Washington, DC: Draft report to the U.S. Environmental Protection Agency.

McKenzie, L.; Hao, W.M.; Richards, G.; Ward, D. 1994 Quantification of major components emitted from smoldering combustion of wood. Atmospheric Environment. 28(20): 3285–3292.

Middleton, P.; Stewart, T.R.; Dennis, R.L. 1983. Modeling human judgments of urban visual air quality. Atmospheric Environment. 17: 1015–1022.

Middleton, P.; Stewart, T.R.; Leary, J. 1985. On the use of human judgment and physical/chemical measurements in visual air quality management. Atmospheric Environment. 12: 1195–1208.

Mobley, Hugh E. 1976. Smoke management—What is it? In: Southern Forest Fire Laboratory Personnel. Southern smoke management guidebook. Gen. Tech. Rep. SE-10. Asheville, NC: U.S. Department of Agriculture, Forest Service, Southeastern Forest Experiment Station: 1–8.

Mobley, Hugh E. 1989. Summary of smoke-related accidents in the South from prescribed fire (1979–1988). Technical Release 90-R-11. American Pulpwood Association.

Morris, R.E.; Yocke, M.A.; Myers, T.C.; Mirabella, V. 1992. Overview of the variable-grid urban airshed model (UAM-V). 85th Annual Meeting of the A&WMA. Pittsburgh, PA: Air & Waste Management Association.

Myer, T.C.; Guthrie, P.D.; Wu, S.Y. 1996. The implementation of a plume-in-grid module in the SARMAP air quality model (SAQM). SYSAPP-96-06, Systems Applications International, Inc. Sacramento, CA: California Air Resources Board, Technical Support Division.

National Interagency Fire Center. 2001a. National fire news: fire season 2000. http://www.nifc.gov/fireinfo/2000/highlights.html. (August 8, 2002).

National Interagency Fire Center. 2001b. Prescribed fire statistics. http://www.nifc.gov/stats/prescribedfirestats.html. (August 8, 2002).

National Interagency Fire Center. 2002. Wildland fires statistics. http://www.nifc.gov/stats/wildlandfirestats.html. (October 5, 2002).

National Research Council. 1993. Protecting visibility in national parks and wilderness areas: committee on haze in National Parks and wilderness areas, National Research Council. Washington, DC: National Academy Press. 446 p.

National Wildfire Coordinating Group. 1985. Smoke management guide. PNW 420-2. NFES 1279. Boise, ID: National Interagency Fire Center, National Interagency Coordinating Group, Prescribed Fire and Fire Effects Working Team. 28 p.

New Mexico Environmental Improvement Board. 1995. Open burning: 20-NMAC2.60. http://www.nmenv.state.nm.us/NMED_regs/aqb/20nmac2_60.html. (August 8, 2002).

New York Daily News. 1999. Florida fire stirs health warnings. April 20; Sports Final Edition; News Section. Page 2.

Nichols, Mary D. 1996. Memorandum dated May 30 to EPA Regional Air Directors. Subject: Areas Affected by PM-10 Natural Events. http://www.epa.gov/ttn/oarpg/t1/memoranda/nepol.pdf.

Odman, M.T.; Russell, A.G. 1991. Multiscale modeling of pollutant transport and chemistry. Journal of Geophysical Research. 96(D4): 7363–7370.

Olson, J.; Prather, M.; Berntsen, T.; Carmichael, G.; Chatfield, R.; Connell, P.; Derwent, R.; Horowitz, L.; Jin, S.; Kanakidou, M.; Kasibhatla, P.; Kotamarthi, R.; Kuhn, M.; Law, K.; Penner, J.; Perliski, L.; Sillman, S.; Stordal, F.; Thompson, A.; Wild, O. 1997. Results from the Intergovernmental Panel on Climate Change photochemical model intercomparison (PhotoComp). Journal of Geophysical Research. 102 (D5): 5979–5991.

Oregon Department of Forestry. 1992. Smoke management program directives, appendix 2. Salem, OR: Oregon Department of Forestry.

Ottmar, R.D.; Reinhardt, T.E.; Anderson, G.; DeHerrera, P.J. [In preparation]. Consume 2.1 user's guide. Manuscript. On file with: Roger D. Ottmar, Pacific Northwest Research Station, Seattle, WA.

Ottmar, Roger D.; Burns, Mary F.; Hall, Janet N.; Hanson, Aaron D. 1993. CONSUME users guide. Gen. Tech. Rep. PNW-GTR-304. Portland, OR: U.S. Department of Agriculture, Forest Service, Pacific Northwest Research Station. 117 p.

Ottmar, Roger D.; Sandberg, David V. 2000. Modification and validation of fuel consumption models for shrub and forested lands in the Southwest, Pacific Northwest, Rockies, Midwest, Southwest, and Alaska. Abstract. Joint fire science program principle investigators meeting; 2000 October 3–5; Reno, NV. http://www.nifc.gov/joint_fire_sci/jointfiresci.html

Ottmar, Roger D.; Vihnanek, Robert E. 2000a. Photo series for major natural fuel types of the United States—phase II. Abstract. Joint fire science program principle investigators meeting; 2000 October 3–5; Reno, NV. http://www.nifc.gov/joint_fire_sci/jointfiresci.html

Peterson, J. 2000. Personal communication. December.

Peterson, Janice L. 1987. Analysis and reduction of the errors of predicting prescribed burn emissions. Seattle: University of Washington. 70 p. Thesis.

Peterson, J.; Ward, D. 1990. An inventory of particulate matter and air toxic emissions from prescribed fire in the United States for 1989. Seattle, WA: U.S. Department of Agriculture, Forest Service, Pacific Northwest Research Station.

Peterson, J.L.; Ward, D. 1992. An inventory of particulate matter and air toxic emissions from prescribed fires in the United States for 1989. Final report. U.S. EPA Office of Air Quality Programs and Standards.

Peterson, Janice L.; Sandberg, David V. 1988. A national PM10 emissions inventory approach for wildfires and prescribed fires. In: Mathai, C.V.; Stonefield, David H., eds. Transactions PM-10 implementation of standards: an APCA/EPA international specialty conference. Pittsburgh, PA: Air Pollution Control Association: 353–371.

Pickering, K.E.; Thompson, A.M.; Scala, J.R.; Tao, W.K.; Simpson, J. 1992. Ozone production potential following convective redistribution of biomass burning emissions. Journal of Atmospheric Chemistry 14(1–4): 297–313.

Pielke, R.A.; Cotton, W.R.; Walko, R.L.; Tremback, C.J.; Lyons, W.A.; Grasso, L.D.; Nicholls, M.E.; Moran, M.D.; Wesley, D.A.; Lee, T.J.; Copeland, J.H. 1992. A comprehensive meteorological modeling system—RAMS. Meteorology and Atmospheric Physics. 49(1–4): 69–91.

Public Law 84-159. Air Pollution Control Act of 1955. Act of July 14, 1955. 42 U.S.C. 7401, et seq. 69 Stat. 322.

Public Law 88-206. Clean Air Act of 1963. Act of December 17, 1963, 77 Stat. 392.

Public Law 90-148. Air Quality Act of 1967. Act of November 1, 1967. 42 U.S.C. 7401. 81 Stat. 485, 501.

Public Law 91-604. Clean Air Act Amendments of 1970. Act of December 31, 1970. 42 USC 1857h-7 et seq.

Public Law 95-95. Clean Air Act as Amended August 1977. 42 U.S.C. s/s 1857 et seq.

Public Law 101-549. Clean Air Act as Amended. November 15, 1990. 104 Stat. 2399.

Radke, L.F.; Lyons, J.H.; Hobbs, P.V.; Hegg, D.A.; Sandberg, D.V.; Ward, D.E. 1990. Airborne monitoring and smoke characterization of prescribed fires on forest lands in western Washington and Oregon. Gen. Tech. Rep. PNW-GTR-251. Portland, OR: U.S. Department of Agriculture, Forest Service, Pacific Northwest Research Station. 81 p.

Reid, J.S.; Hobbs, P.V.; Ferek, R.J.; Blake, D.R.; Martins, J.V.; Dunlap, M.R.; Liousse, C. 1998. Physical, chemical and optical properties of regional hazes dominated by smoke in Brazil. Journal of Geophysical Research (SCAR-B Special Issue). 103: 32,059–32,080.

Reinhardt, Elizabeth D.; Keane, Robert E. 2000. A national fire effects prediction model—revision of FOFEM. Abstract. Joint fire science program principle investigators meeting; 2000 October 3–5; Reno, NV. http://www.nifc.gov/joint_fire_sci/jointfiresci.html

Reinhardt, Elizabeth D.; Keane, Robert E.; Brown, James K. 1997. First Order Fire Effects Model: FOFEM 4.0, users guide. Gen. Tech. Rep. INT-GTR-344. Ogden, UT: U.S. Department of Agriculture, Forest Service, Intermountain Research Station. 65 p.

Reinhardt, T.E. 2000. Effects of smoke on wildland firefighters. Seattle, WA: URS/Radian International. April 2000.

Reinhardt, T.E.; Ottmar, R.D. 2000. Smoke exposure at western wildfires. Res. Pap. PNW-RP-525. Portland, OR: U.S. Department of Agriculture, Forest Service, Pacific Northwest Research Station. 72 p.

Reinhardt, Tim E.; Ottmar, Roger D.; Hallett, Michael J. 1999. Guide to monitoring smoke exposure of wildland firefighters. Gen. Tech. Rep. PNW-GTR-448. U.S. Department of Agriculture, Forest Service, Pacific Northwest Research Station. 15 p.

Reinhardt, Timothy E.; Ottmar, Roger D. 1997. Smoke exposure among wildland firefighters: a review and discussion of current literature. Gen. Tech. Rep. PNW-GTR-373. U.S. Department of Agriculture, Forest Service, Pacific Northwest Research Station. 61 p.

Reinhardt, Timothy E.; Ottmar, Roger D.; Hanneman, Andrew J.S. 2000. Smoke exposure among firefighters at prescribed burns in the Pacific Northwest. Res. Pap. PNW-RP-526. U.S. Department of Agriculture, Forest Service, Pacific Northwest Research Station. 45 p.

Reisner, J.; Wynne, S.; Margolin, L.; Linn, R. 2000. Coupled atmospheric-fire modeling employing the method of averages. Monthly Weather Review. 128(10): 3683–3691.

Richardson, J.L.; Fishman, J.; Gregory, G.L. 1991. Ozone budget over the Amazon: regional effects from biomass burning emissions. Journal of Geophysical Research. 96(D7): 13,073–13,087.

Riebau, A.R.; Fox, D.G.; Sestak, M.L.; Daily, B.; Archer, S.F. 1988. Simple approach smoke estimation model. Atmospheric Environment. 22(4): 783–788.

Roger, C.F.; Hudson, J.G.; Zielinska, B.; Tanner, R.L.; Hallett, J.; Watson, J.G. 1991. Cloud condensation from biomass burning. In: Levin, J.S., ed. Global burning: atmospheric, climatic, and biospheric implications. Cambridge, MA: MIT Press.

Ross, D. 1988. Effects of visual air quality on visitor experience. In: Air quality in the national parks: A summary of findings from the National Park Service Air Quality Research and Monitoring Program. Natural Resources Report 88-1. Chapter 3.

Ross, D.G.; Smith, I.N.; Mannis, P.C.; Fox, D.G. 1988. Diagnostic wind field modeling for complex terrain: model development and testing. Journal of Applied Meteorology. 27: 785–796.

Rothermel, R.C. 1972. A mathematical model for predicting fire spread in wildland fuels. Res. Pap. INT-115. Ogden, UT: U.S. Department of Agriculture, Forest Service, Intermountain Forest and Range Experiment Station. 40 p.

Rothwell, T.; Sharkey, B. 1995. The effect of an air-purifying respirator on performance of upper body work. In: Health hazards of smoke: fall 1995. U.S. Department of Agriculture, Forest Service, Missoula Technology and Development Center: 56–65.

Sandberg, D.V. 1980. Duff reduction by prescribed underburning in Douglas-fir. Res. Pap. PNW-272. Portland, OR: U.S. Department of Agriculture, Forest Service, Pacific Northwest Forest and Range Experiment Station. 18 p.

Sandberg, David V. 2000. Implementation of an improved Emission Production Model. Abstract. Joint fire science program principle investigators meeting; 2000 October 3–5; Reno, NV. http://www.nifc.gov/joint_fire_sci/jointfiresci.html.

Sandberg, David V. 2002. Personal communications. Supervisory Research Biologist. U.S. Department of Agriculture, Forest Service, Pacific Northwest Research Station. October.

Sandberg, D.V.; Dost, F.N. 1990. Effects of prescribed fire on air quality and human health. In: Wasltad, J.W.; Radosevich, S.R.; Sandberg, D.V., eds. Natural and prescribed fire in Pacific Northwest forests. Corvallis: Oregon State University Press: 191–218.

Sandberg, David V.; Hardy, Colin C.; Ottmar, Roger D.; Snell, J.A. Kendall; Acheson, Ann; Peterson, Janice L.; Seamon, Paula; Lahm, Peter; Wade, Dale. 1999. National strategic plan: Modeling and data systems for wildland fire and air quality. U.S. Department of Agriculture, Forest Service, Pacific Northwest Research Station. 60 p.

Sandberg, David V.; Ottmar, R.D.; Cushon, G.H. 2001. Characterizing fuels in the 21st century. International Journal of Wildland Fire. 10: 1–7.

Sandberg, D.V.; Peterson, J.L. 1984. A source strength model for prescribed fires in coniferous logging slash. Annual Meeting. Air Pollution Control Association, Pacific Northwest Section. Reprint #84.20. Portland, OR: U.S. Department of Agriculture, Forest Service, Pacific Northwest Research Station. 10 p.

Sandberg, D.V.; Pierovich, J.M.; Fox, D.G.; Ross, E.W. 1979. Effects of fire on air: a state-of-knowledge review. Gen. Tech. Rep. WO-9. U.S. Department of Agriculture, Forest Service. 40 p.

Schimel, D.S. 1995. Terrestrial ecosystems and the global carbon cycle. Global Change Biology. 1: 77–71.

Schroeder, W.H.; Lane, D.A. 1988. The fate of toxic airborne pollutants. Environmental Science and Technology. 22(3): 240–246.

Schulze, William D.; Brookshire, David S.; Walther, Eric G.; MacFarland, Karen Kelley; Thayer, Mark A.; Whitworth, Regan L.; Ben-David, Shaul; Malm, William; Molenar, John. 1983. The economic benefits of preserving visibility in the national parklands of the Southwest. Natural Resources Journal. 23: 149–173.

Schwartz, Joel; Dockery, Douglas W.; Neas, Lucas M. 1996. Is daily mortality associated specifically with fine particles? Journal of the Air & Waste Management Association. 46: 927–939.

Scire, J.; Robe, F.R.; Fernau, M.E.; Yamartino, R.J. 2000a. A user's guide for CALMET meteorological model. Concord, MA: Earth Tech, Inc. 332 p.

Scire, J.; Strimaitis, D.G.; Yamartino R.J.; Xiaomong, Zhang. 2000b. A user's guide for CALPUFF dispersion model (Version 5). Concord, MA: Earth Tech, Inc. 512 p.

Seigneur, C.; Tesche, T.W.; Roth, P.M.; Liu, M.K. 1983. On the treatment of point source emissions in urban air quality. Atmospheric Environment. 17(9): 1655–1676.

Sestak, M.L.; Riebau, A.R. 1988. SASEM, Simple approach smoke estimation model. U.S. Bureau of Land Management, Technical Note 382. 31 p.

Sestak, M.L.; Marlatt, W.E.; Riebau, A.R. 1988. Unpublished draft. VALBOX: ventilated valley box model. U.S. Bureau of Land Management. 32 p.

Sharkey, Brian, ed. 1997. Health hazards of smoke: recommendations of the April 1997 Consensus Conference. Tech. Rep. 9751-2836-MTDC. Missoula, MT: U.S. Department of Agriculture, Forest Service, Missoula Technology and Development Center. 84 p.

USDA Forest Service Gen. Tech. Rep. RMRS-GTR-42-vol. 5. 2002

73

Shelby, B.; Speaker, R.W. 1990. Public attitudes and perceptions about prescribed burning. In: Wasltad, J.W.; Radosevich, S.R.; Sandberg, D.V., eds. Natural and prescribed fire in Pacific Northwest forests. Corvallis: Oregon State University Press: 253–260.

Sisler, James F.; Malm, William C.; Gebhart, Kristi A. 1996. Spatial and seasonal patterns and long term variability of the composition of the haze in the United States: an analysis of data from the IMPROVE network. Cooperative Institute for Research in the Atmosphere, Colorado State University. ISSN: 0737-5352-32. [Pages unknown].

Southern Forest Fire Laboratory Personnel. 1976. Southern forestry smoke management guidebook. Gen. Tech. Rep. SE-10. Asheville, NC: U.S. Department of Agriculture, Forest Service, Southeastern Forest Experiment Station. 140 p.

Standley, L. J.; Simoneit, B. R. T. 1987. Characterization of extractable plant wax, resin and thermally matured components in smoke particles from prescribed burns. Environmental Science and Technology. 21: 163–169.

Stith, J.L.; Radke, F.L.; Hobbs, P.V. 1981. Particle emissions and the production of ozone and nitrogen oxides from the burning of forest slash. Atmospheric Environment. 7: 73–82.

Systems Applications International. 2002. User's guide to the regulatory modeling system for aerosols and deposition (REMSAD). SYSAPP98-96/42r2. Contract 68D30032. U.S. Environmental Protection Agency. http://remsad.saintl.com/ or http://www.epa.gov/scram001/

Thompson, A.M.; Pickering, K.E.; McNamera, D.P.; Schoeberl, M.R.; Hudson, R.D.; Kim, J.H.; Browell, E.V.; Kirchoff, V.W.J.H.; Ngana, D. 1996. Where did tropospheric ozone over southern Africa and the tropical Atlantic come from in October 1992? Insights from TOMS, GTE TRACE A, and SAFARI 1992. Journal of Geophysical Research. 101(D19):24,251–24,278.

Trent, Andy; Davies, Mary Ann; Fisher, Rich; Thistle, Harold; Babbitt, Ronald. 2000. Evaluation of optical instruments for real-time continuous monitoring of smoke particles. Tech. Rep. 9925-2806-MTDC. Missoula, MT: U.S. Department of Agriculture, Forest Service, Technology and Development Program. 38 p.

Trent, Andy; Thistle, Harold; Fisher, Rich; Babbitt, Ronald; Holland-Sears, Andria. 1999. Laboratory evaluation of two optical instruments for real-time particulate monitoring of smoke. Tech. Rep. 9925-2806-MTDC. Missoula, MT: U.S. Department of Agriculture, Forest Service, Technology and Development Program. 38 p.

Trijonis, J.; Charlson, R.; Husar, R.; Malm, W.C.; Pitchford, M.; White, W. 1991. Visibility: existing and historical conditions—causes and effects. In: Acid deposition: state of science and technology: Report 24. National Acid Precipitation Assessment Program. Washington, DC: Government Printing Office. [Pages unknown].

U.S. Code Title 42, Chapter 85. Air Pollution Prevention and Control, As Amended.

USA Today. 1999. Fires blaze in six Western states. http://www.usatoday.com/weather/news/1999/w827fire.htm. August 27, 1999.

U.S. Department of Agriculture. 1997. Course to the future: the RPA program. http://www.fs.fed.us/land/RPA/chp3sec1.htm. April 1997.

U.S. Department of Agriculture and U.S. Department of the Interior. 2000. A report to the President in response to the wildfires of 2000. http://www.fireplan.gov/president.cfm. September 2002.

U.S. Department of the Interior; U.S. Department of Agriculture. 1995. Federal wildland fire management policy and program review. Final report. Boise, ID: Bureau of Land Management. 45 p.

U.S. Department of the Interior; U.S. Department of Agriculture; Department of Energy; [and others]. 2001. Review and update of the 1995 federal wildland fire management policy. Boise, ID: Bureau of Land Management. 78 p.

U.S. Environmental Protection Agency. 1972. Compilation of air pollutant emission factors. OAP Publ. AP-42. Research Triangle Park, NC: U.S. Environmental Protection Agency.

U.S. Environmental Protection Agency. 1990. Air quality criteria for particulate matter: Volume II of III. EPA/600/P-95/00ibF. Washington, DC: U.S. Environmental Protection Agency, Office of Research and Development, National Center for Environmental Assessment: 8-82–8-89.

U.S. Environmental Protection Agency. 1992a. Prescribed burning background document and technical information document for prescribed burning best available control measures. EPA-450/2-92-003. Office of Air Quality Planning and Standards. September.

U.S. Environmental Protection Agency. 1992b. Prevention of air pollution emergency episodes. 40 CFR 51 Appendix L.

U.S. Environmental Protection Agency. 1996. Review of the national ambient air quality standard for particulate matter: Policy assessment of scientific and technical information. EPA-452\R-96-013. Washington, DC: U.S Environmental Protection Agency, Office of Air Quality Planning and Standards. July.

U.S. Environmental Protection Agency. 1998. Interim air quality policy on wildland and prescribed fires.

U.S. Environmental Protection Agency, Office of Air Quality Planning and Standards. April 23. 29 p.

U.S. Environmental Protection Agency. 1999. Guideline for reporting of daily air quality – air quality index (AQI). EPA-454/R-99-010. Research Triangle Park, NC: Office of Air Quality Planning and Standards. 25 p.

U.S. Environmental Protection Agency. 2000a. Compilation of air pollutant emission factors AP-42, fifth edition, volume I: stationary point and area sources. Research Triangle Park, NC: U.S. Environmental Protection Agency. January 1995–September 2000. AP-42 reference.

U.S. Environmental Protection Agency. 2000b. National ambient air quality standards (NAAQS). http://www.epa.gov/airs/criteria.html. 5 December 2000.

U.S. Environmental Protection Agency. 2000c. Wildland fire issues group. http://www.epa.gov/ttncaaa1/faca/fa08.html. December 5, 2000.

U.S. Environmental Protection Agency. 2002. Models-3 Air Quality Modeling System. http://www.epa.gov/asmdnerl/models3/July 2002.

Venkatram, A. 1988. Topics in applied dispersion modeling. In: Venkatram, A.; Wyngaard, eds. Lectures on air pollution modeling. Boston, MA: American Meteorological Society: 267–324.

Wade, Dale D.; Lunsford, James D. 1988. A guide to prescribed fire in southern forests. Tech. Publ. R8-TP-11. NFES 2108. U.S. Department of Agriculture, Forest Service, Southern Region. Boise, ID: Publication Management System.

Walcek, C.J. 2002. Effects of wind shear on pollution dispersion. Atmospheric Environment. 36: 511–517.

Walton, W.D.; McGrattan, K.B.; Mullin, J.V. 1996. ALOFT-PC: A smoke plume trajectory model for personal computers. In: Proceedings 19th Arctic and Marine Oilspill Programme (AMOP) Technical Seminar: Volume 2. Ottawa, Ontario, Canada: Environment Canada: 987–997.

Ward, D.E.; Hardy, C.C.; Sandberg, D.V.; Reinhardt, T.E. 1989. Part III-emissions characterization. In; Sandberg, D.V.; Ward, D.E.; Ottmar, R.D., comp. eds. Mitigation of prescribed fire atmospheric pollution through increased utilization of hardwoods, piled residues, and long-needled conifers. Final report. U.S. DOE, EPA. Seattle, WA: U.S. Department of Agriculture, Forest Service, Pacific Northwest Research Station.

Ward, D.E.; McMahon, C.K.; Johansen, R.W. 1976. An update on particulate emissions from forest fires. In: Transactions of the 69th annual meeting of the Air Pollution Control Association.

Ward, Darold E.; Hardy, Colin C. 1991. Smoke emissions from wildland fires. Environmental International. 17: 117–134.

Washington Department of Natural Resources. 1993. Smoke management program, Appendix 5. Olympia, WA.

Watson, J.W. 1997. Review and comment on observational models—the use of chemical mass balance methods. In: Proceedings of the EPA source attribution workshop. Research Triangle Park, NC: U.S. Environmental Protection Agency.

Watson, John G.; Robinson, Norman F.; Chow, Judith C.; [and others]. 1990. CMB7 user's manual. Receptor model technical series, volume III (1989 revision). EPA-450/4-90-004. Research Triangle Park, NC: U.S. Environmental Protection Agency, Office of Air and Radiation, Office of Air Quality Planning and Standards.

WESTAR. 1995. Wildfire emergency action plans draft report. Portland, OR: Western States Air Resources Council.

74

USDA Forest Service Gen. Tech. Rep. RMRS-GTR-42-vol. 5. 2002

Western Governors' Association. 2001. A collaborative approach for reducing wildland fire risks to communities and the environment: 10-year comprehensive strategy. http://www.westgov.org/wga/initiatives/fire/final_fire_rpt.pdf. August 2001.

Williamson, Samuel J. 1973. Fundamentals of air pollutions. Reading, MA: Addison-Wesley Publishing Company. 472 p.

Wofsy, S.C.; Sachse, G.W.; Sachse, G.L.; Blake, D.L.; Bradshaw, J.D.; Singh, H.B.; Barrick, J.A.; Harriss, R.C.; Talbot, R.W.; Shipman, M.A.; Browell, E.V.; Jacob, D.J.; Logan, J.A. 1992. Atmospheric chemistry in the Arctic and Subarctic: influence of natural fires, industrial emissions, and stratospheric inputs. Journal of Geophysical Research. 97: 16,731–16,746.

Wotawa, G.; Trainer, M. 2000. The influence of Canadian forest fires on pollutant concentrations in the United States. Science. 288: 324–328.

Yamate, G.; Stockham, J.; Vatavuk, W.; Mann, C. 1975. An inventory of emissions from forest wildfires, forest managed burns and agricultural burns. In: Transactions of the 68th annual meeting of the Air Pollution Control Association. June 15–20, 1975.

Yocom, J.E.; Upham, J.B. 1977. Effects of economic materials and structures. In: Stern, Arthur C., ed. Air pollution: The effects of air pollution. 3rd ed. New York, NY: Academic Press, Inc.: 93–94.

Yokelson, R.J.; Griffith, D.W.T.; Ward, D.E. 1996. Open-path Fourier transform infrared studies of large scale laboratory biomass fires. Journal of Geophysical Research. 101(D15): 20,167–21,080.

Yokelson, R.J.; Ward, D.E.; Susott, R.A.; Reardon, J.; Griffith, D.W.T. 1997. Emissions from smoldering combustion of biomass measured by open-path Fourier transform infrared spectroscopy. Journal of Geophysical Research. 102(D15): 18,865–18,877.

USDA Forest Service Gen. Tech. Rep. RMRS-GTR-42-vol. 5. 2002

75

Index

H

HAP. *See* air pollutants: hazardous air pollutants (HAP)
health 2, 4, 6, 11, 16, 20, 24, 35, 45, 55, 56, 57, 63, 66
   community 7
   effects 1, 11, 45, 55, 56, 57, 65
   firefighter 65
   human 6, 9, 11, 16
   public 6, 7, 20
   risks 5, 45, 57, 64
HIGRAD. *See* models
hydrocarbons. *See* air pollutants
Hysplit. *See* models

I

Idaho. *See* United States of America: Idaho
Illinois. *See* United States of America: Illinois
IMPROVE 48, 49, 51
Iowa. *See* United States of America: Iowa

J

Joint Fire Science Program (JFSP) 2, 6, 64, 67

L

laws and regulations
   Clean Air Act 1, 4, 6, 9, 10, 11, 12, 13, 15, 16, 49, 58, 59
      roles and responsibilities 10
   general conformity 10, 11
   National Environmental Policy Act (NEPA) 4
liability. *See* safety
Louisiana. *See* United States of America: Louisiana

M

Massachussetts. *See* United States of America: Massachusetts
meteorology 4, 10, 60, 64
Michigan. *See* United States of America: Michigan
Midwest Regional Planning Organization. *See* regional
      planning organizations
Mississippi. *See* United States of America: Mississippi
model types
   chemical mass balance (CMB) 51
   dispersion 32, 35, 36, 37, 38, 40, 50, 64, 65, 66
   grid 32, 37, 39, 40
   particle 39
   plume 37, 38, 39, 40, 60
   puff 37, 39, 40
   scavenging 38
   transport 35, 37
models
   ALOFT-FT 39
   BEHAVE 29
   BURNUP 29, 33, 36, 37
   Cal/PFIRS 48
   CALMET 38, 64
   Calpuff 36, 39, 40, 64
   CMAQ 32
   Consume 31, 32
   Emission production model (EPM) 29, 32, 36, 37
   FARSITE 29, 33, 36
   FASTRACS 48

FIRETEC 29
FOFEM 31, 32
HIGRAD 29
Hysplit 39, 40
MM5 38
Models-3/CMAQ 32, 36, 38, 40, 41
NFSpuff 39, 40
NUATMOS 38
PB-Piedmont 39, 61
RAMs 38
REMSAD 40
SASEM 36, 37, 39, 40, 64
TSARS+ 38
VSMOKE 36, 37, 39, 60
Models-3/CMAQ. *See* models
monitoring 3, 4, 5, 11, 15, 17, 25, 26, 45, 48, 49, 50,
      51, 52, 53, 57, 63, 64, 66
Montana. *See* United States of America

N

NAAQS 4, 5. *See* national ambient air quality standards
      (NAAQS)
national ambient air quality standards
      (NAAQS) 4, 5, 10, 11, 16, 17, 19, 24, 25, 48, 49, 60, 64
National Fire Plan 7, 67
National Interagency Fire Center (NIFC) 6, 24
National Wildfire Coordinating Group (NWCG) 3, 4
NEAP. *See* smoke management planning: natural events
      action plan (NEAP)
NEPA. *See* laws and regulations: National Environmental
      Policy Act (NEPA)
Nevada. *See* United States of America: Nevada
New Mexico. *See* United States of America: New Mexico
New York. *See* United States of America: New York
NFSpuff. *See* models
NIFC. *See* National Interagency Fire Center (NIFC)
nonattainment 10, 11, 16, 24, 25, 64
North Carolina. *See* United States of America: North Carolina
NWCG. *See* National Wildfire Coordinating Group (NWCG)

O

Oregon. *See* United States of America: Oregon
OTC. *See* regional planning organizations
ozone 5, 10, 11, 17, 39, 40, 41, 42, 43, 45, 49, 65, 66
Ozone Transport Commission. *See* regional planning
      organizations

P

Pacific Northwest. *See* United States of America: Pacific
      Northwest
particulate matter 4, 5, 11, 20, 24, 32, 48, 49, 50, 55,
      56, 59, 64
particulates 9, 10
PB-Piedmont. *See* models
PFIRS. *See* smoke management programs: Prescribed Fire
      Incident Reporting System
photo series 28
Piedmont. *See* United States of America: Piedmont
plume 17, 23, 24, 27, 29, 36, 37, 38, 39, 40, 41, 42, 43,
      50, 58, 59, 61

chemistry   38, 39, 42
plume rise   35, 36, 37, 39, 40, 66
prevention of significant deterioration (PSD)   10, 12, 13

R

RAMs. *See* models
regional haze   4, 5, 10, 13, 14, 15, 17, 24, 26, 36, 39,
    40, 41, 43, 45, 46, 48, 51, 53, 64, 67
regional planning organizations   14, 40
    Central States Regional Air Partnership   14
    Midwest Regional Planning Organizationrship   14
    Ozone Transport Commission   41
    Visibility Improvement State and Tribal Associatio   14
REMSAD. *See* models
research questions   1, 2, 4, 7, 10, 17, 31, 37, 40, 42,
    43, 47, 48, 57, 59, 63, 64, 65, 66, 67
Rocky Mountain. *See* United States of America: Rocky
    Mountain
Russia. *See* World: Russia

S

safety   1, 4, 6, 16, 63, 66
    firefighter   7
    highway   23, 58, 60, 66
    liability   5, 25
    public   6, 25, 60
    roadway   23, 60
SASEM. *See* models
scavenging   36, 37, 38, 40
SIP. *See* smoke management programs: State
    implementation plan (SIP)
smoke
    dispersion   4, 15, 17, 26, 36
    prescribed fire   25, 52, 60
    reduction. *See* emission(s): reduction
    transport   6, 14, 25, 26
    wildland fire   49, 52, 56
smoke management guidance
    smoke management guide for prescribed and wildland
      fire   2, 3, 4, 5, 17
    southern smoke management guidebook   2, 6
smoke management guide for prescribed and wildland
    fire. *see* smoke management guidance
smoke management planning   16
    enhanced smoke management plan   26
    natural events action plan (NEAP)   16, 24
smoke management
    programs   4, 5, 15, 16, 17, 19, 24, 25, 58
    Prescribed Fire Incident Reporting System   26
    State implementation plan
      (SIP)   5, 10, 11, 16, 17, 40, 46
    Tribal implementation plan (TIP)   10, 40
source strength   2, 32, 33, 36, 39, 40, 52, 64, 65, 66
Southern Smoke Management Guidebook. *See* smoke
    management guidance
South America. *See* World: South America
South Carolina. *See* United States of America: South
    Carolina

T

Tennessee. *See* United States of America: Tennessee
Texas. *See* United States of America: Texas
TIP. *See* smoke management programs: Tribal implementa-
    tion plan (TIP)

U

U.S. Environmental Protection Agency   4, 5, 7, 10, 12,
    14, 16, 17, 19, 20, 24, 25, 39, 41, 47, 48, 55, 57, 58, 59
United States of America
    Alabama   25
    Alaska   19, 25, 40, 42, 63
    Appalachian Mountains   61
    Arizona   24, 25, 39, 46
    Arkansas   25
    California   15, 19, 20, 24, 25, 26, 46, 48, 58
      Central Valley   20
      San Joaquin Valley   26
      southern   58
    Colorado   25, 39, 46
    Colorado Plateau   59
    Eastern Seaboard   20
    Eastern U.S.   51
    Florida   19, 20, 25, 46, 60
    Georgia   24, 25
    Grand Canyon   59
    Idaho   20, 24, 46
      Salmon   20
    Illinois   25
    Intermountain West   19
    Iowa   58
    Louisiana   25
    Massachusetts   25
    Michigan   46
    Midwestern States   41
    Mississippi   25, 60
    Montana   20, 24, 46
    Montana/Idaho   15, 25
    Nevada   20, 25, 26, 46
      Reno   20
    New Mexico   20, 25, 39, 46, 58
    New York   25
    North Carolina   25
    Northeastern   15
    Northwestern U.S.   51
    Ohio   60
    Ohio River   60
    Oregon   12, 20, 24, 25, 46, 48, 58
      southern   20
    Pacific Coast   20
    Pacific Northwest   15, 48, 51
    Piedmont   61
    Rocky Mountain   15
    Rocky Mountain States   20
    South Carolina   25, 46
    Southeastern   15, 19, 20, 23, 24, 60
    Southern U.S.   60
    Southwest   20, 51
    Tennessee   25, 41, 46

78

USDA Forest Service Gen. Tech. Rep. RMRS-GTR-42-vol. 5. 2002

Texas   24, 25, 60
Utah   25, 46
Virginia   25, 60
Washington   20, 24, 25, 46, 48, 58
Western Regional Air Partnership
   (WRAP)   14, 26, 40, 41, 67
Western U.S.   15, 19
Wyoming   25, 39
Yellowstone National Park   20
Utah. *See* United States of America: Utah

V

Virginia. *See* United States of America: Virginia
visibility   2, 4, 6, 7, 10, 13, 14, 15, 19, 20, 24, 25, 26, 35,
        39, 41, 45, 49, 50, 58, 59, 60, 64, 65, 66
   Class I
      areas   10, 12, 14, 15, 24, 25, 47, 49, 50, 58, 59, 65
   visibility impairment and reduction   1, 20, 46, 58, 60
VISTAS. *See* regional planning organizations
VSMOKE. *See* models

W

Washington. *See* United States of America: Washington
welfare   2, 4, 6, 9, 11, 16, 35, 45, 55, 58, 66
Western Regional Air Partnership (WRAP)   14, 67
World
   Africa
      tropical   24
   Asia
      tropical   24
   Canada   20, 25, 42, 49
   Russia   25
   South America
      tropical   25
WRAP. *See* Western Regional Air Partnership (WRAP)
Wyoming. *See* United States of America: Wyoming

Y

Yellowstone National Park. *See* United States of America:
   Yellowstone National Park

RMRS
ROCKY MOUNTAIN RESEARCH STATION

The Rocky Mountain Research Station develops scientific information and technology to improve management, protection, and use of the forests and rangelands. Research is designed to meet the needs of National Forest managers, Federal and State agencies, public and private organizations, academic institutions, industry, and individuals.

Studies accelerate solutions to problems involving ecosystems, range, forests, water, recreation, fire, resource inventory, land reclamation, community sustainability, forest engineering technology, multiple use economics, wildlife and fish habitat, and forest insects and diseases. Studies are conducted cooperatively, and applications may be found worldwide.

Research Locations

Flagstaff, Arizona	Reno, Nevada
Fort Collins, Colorado*	Albuquerque, New Mexico
Boise, Idaho	Rapid City, South Dakota
Moscow, Idaho	Logan, Utah
Bozeman, Montana	Ogden, Utah
Missoula, Montana	Provo, Utah
Lincoln, Nebraska	Laramie, Wyoming

*Station Headquarters, Natural Resources Research Center, 2150 Centre Avenue, Building A, Fort Collins, CO 80526

www.ingramcontent.com/pod-product-compliance
Lightning Source LLC
Chambersburg PA
CBHW081612220526
45468CB00010B/2846

* 9 7 8 1 9 7 3 8 0 8 0 3 9 *